AF886343

Decolorization by Thanatephorus Cucumeris Dec 1

Authored by

Makoto Shoda
Emeritus of Tokyo Institute of Technology
316,1-4-2 Shin-Ishikawa, Aobaku
Yokohama, Japan

Decolorization by Thanatephorus Cucumeris Dec 1

Author: Makoto Shoda

ISBN (Online): 978-981-5256-89-5

ISBN (Print): 978-981-5256-90-1

ISBN (Paperback): 978-981-5256-91-8

© 2024, Bentham Books imprint.

Published by Bentham Science Publishers Pte. Ltd. Singapore. All Rights Reserved.

First published in 2024.

BENTHAM SCIENCE PUBLISHERS LTD.

End User License Agreement (for non-institutional, personal use)

This is an agreement between you and Bentham Science Publishers Ltd. Please read this License Agreement carefully before using the ebook/echapter/ejournal (**"Work"**). Your use of the Work constitutes your agreement to the terms and conditions set forth in this License Agreement. If you do not agree to these terms and conditions then you should not use the Work.

Bentham Science Publishers agrees to grant you a non-exclusive, non-transferable limited license to use the Work subject to and in accordance with the following terms and conditions. This License Agreement is for non-library, personal use only. For a library / institutional / multi user license in respect of the Work, please contact: permission@benthamscience.net.

Usage Rules:

1. All rights reserved: The Work is the subject of copyright and Bentham Science Publishers either owns the Work (and the copyright in it) or is licensed to distribute the Work. You shall not copy, reproduce, modify, remove, delete, augment, add to, publish, transmit, sell, resell, create derivative works from, or in any way exploit the Work or make the Work available for others to do any of the same, in any form or by any means, in whole or in part, in each case without the prior written permission of Bentham Science Publishers, unless stated otherwise in this License Agreement.
2. You may download a copy of the Work on one occasion to one personal computer (including tablet, laptop, desktop, or other such devices). You may make one back-up copy of the Work to avoid losing it.
3. The unauthorised use or distribution of copyrighted or other proprietary content is illegal and could subject you to liability for substantial money damages. You will be liable for any damage resulting from your misuse of the Work or any violation of this License Agreement, including any infringement by you of copyrights or proprietary rights.

Disclaimer:

Bentham Science Publishers does not guarantee that the information in the Work is error-free, or warrant that it will meet your requirements or that access to the Work will be uninterrupted or error-free. The Work is provided "as is" without warranty of any kind, either express or implied or statutory, including, without limitation, implied warranties of merchantability and fitness for a particular purpose. The entire risk as to the results and performance of the Work is assumed by you. No responsibility is assumed by Bentham Science Publishers, its staff, editors and/or authors for any injury and/or damage to persons or property as a matter of products liability, negligence or otherwise, or from any use or operation of any methods, products instruction, advertisements or ideas contained in the Work.

Limitation of Liability:

In no event will Bentham Science Publishers, its staff, editors and/or authors, be liable for any damages, including, without limitation, special, incidental and/or consequential damages and/or damages for lost data and/or profits arising out of (whether directly or indirectly) the use or inability to use the Work. The entire liability of Bentham Science Publishers shall be limited to the amount actually paid by you for the Work.

General:

1. Any dispute or claim arising out of or in connection with this License Agreement or the Work (including non-contractual disputes or claims) will be governed by and construed in accordance with the laws of Singapore. Each party agrees that the courts of the state of Singapore shall have exclusive jurisdiction to settle any dispute or claim arising out of or in connection with this License Agreement or the Work (including non-contractual disputes or claims).
2. Your rights under this License Agreement will automatically terminate without notice and without the

need for a court order if at any point you breach any terms of this License Agreement. In no event will any delay or failure by Bentham Science Publishers in enforcing your compliance with this License Agreement constitute a waiver of any of its rights.

3. You acknowledge that you have read this License Agreement, and agree to be bound by its terms and conditions. To the extent that any other terms and conditions presented on any website of Bentham Science Publishers conflict with, or are inconsistent with, the terms and conditions set out in this License Agreement, you acknowledge that the terms and conditions set out in this License Agreement shall prevail.

Bentham Science Publishers Pte. Ltd.
80 Robinson Road #02-00
Singapore 068898
Singapore
Email: subscriptions@benthamscience.net

BENTHAM
SCIENCE

CONTENTS

PREFACE

This book summarizes the results of more than 20 years of research on a fungus, Dec 1, which was isolated in my laboratory, as a candidate to decolorize colored substances that are visible xenobiotic pollutants and cause serious environmental problems.

Various studies have attempted to improve or change environments using biological activity, but practical examples of this technique are rare. This is mainly because of the lack of basic information and the lack of optimization in reactor systems. In this book, one possibility of using a single microorganism to resolve environmental issues related to colored substances is demonstrated.

This book is composed of six chapters, and various aspects of isolated Dec 1, including its scientific analysis and the optimization of the engineered production of enzymes responsible for decolorization, are provided in it.

The features and characteristics of Dec 1 included in this book are as follows:

1. Dec 1 has been isolated as a fungus for dye decolorization. It has been finally identified as the non-white rot fungus, *Thanatephorus cucmerins* Dec 1.

2. The decolorization spectrum of Dec 1 is vast, indicating that Dec 1 produces multiple enzymes that are responsible for decolorization.

3. New peroxidase DyP, Aryl Alcohol Oxidase (AAO), Manganese Peroxidase (MnP), and Versatile Peroxidase (TcVP1) have been purified from Dec 1 and characterized.

4. The gene, dyp, has been cloned from Dec 1 and expressed in the fungus, Aspergillus oryzae.

5. Recombinant DyP (rDyP) has been shown to be almost identical to DyP.

6. The unique characteristics of DyP have been verified using crystallization and X-ray analysis.

7. The immobilization of rDyP has been found to be successful only when using a new mesoporous material as a carrier.

8. The enhanced production of DyP has been attempted using liquid culture, Solid-state Culture (SSC), repeated-batch culture, and fed-batch culture.

9. To overcome the problems involved in these culture methods, an Air Membrane Surface (AMS) reactor has been introduced, and increased rDyP production has been confirmed.

10. The final level of rDyP production has been found to be more than a half million-fold higher than the original level of DyP production by *T. cucumeris* Dec 1.

11. Dec 1 has demonstrated efficient decolorization of molasses waste, kraft pulp bleaching effluent, and oxygen-delignified bleaching effluent.

12. As substrates for Dec 1 growth, complex media, including rice bran powder, wheat bran

powder, and molasses, have been used, and the advantages of each medium have been described.

I believe that this book will provide researchers in this field with a useful resource to support current knowledge on biological decolorization and will provide students with a logical and practical scheme for approaching colored substance treatment. During my time as the Director of the Resources Recycling Process Laboratory, at the Chemical Resources Laboratory, Tokyo Institute of Technology, many researchers and graduate and undergraduate students at the campus and in private companies have helped to accomplish these results. I am indebted to the following researchers for giving me the opportunity to study Dec 1 and for offering many valuable suggestions: Drs. Y. Sugano, M. Hirai, T. Sato, M. Iwamoto, T.H. Lee, N. Uematsu, and J. Sugiura. I am also grateful to the following graduate and undergraduate students: S.J. Kim, M. Shakeri, N. Shintani, T. Shimokawa, K. Sasaki, Y. Matsushima, A. Ichiyanagi, R. Muramatu, R. Sasaki, N. Suzuki, and C. Matsuo.

I especially thank Professors T. Imanaka, Professor Emeritus of Kyoto University, and J. Takahashi, Professor Emeritus of Obihiro University of Agriculture and Veterinary Medicine, for their cooperation on this book.

Makoto Shoda
Emeritus of Tokyo Institute of Technology
316,1-4-2 Shin-Ishikawa, Aobaku
Yokohama, Japan

CHAPTER 1

Introduction

Several colored effluents, including dyes, molasses, and pulp bleaching effluents, are released into environments, and efficient treatments of these colored substances are urgent because severe regulation has been established worldwide. Conventional treatment methods have several disadvantages, such as energy consumption or the risk of treated products. On the other hand, biological treatment has advantages over conventional chemical and physical treatment because the products after treatment are relatively safer and the utilization of energy is smaller than other methods. However, in biological methods, finding efficient microorganisms and utilizing them under optimal conditions are key challenges.

A fungus, strain Dec 1 was isolated in our laboratory, exhibiting a wide spectrum of degradability for colored substances [1]. This fungus was initially identified as *Geotrichum candidum* Dec 1, but finally reidentified as *Thanatephorus cucumeris* Dec 1. In this book, the characteristics of Dec 1 have been described not only from the basic biochemical points, but also from an engineering point of view associated with its application for the decolorization of dyes, molasses, and pulp bleaching effluents.

More than 5×10^4 tons of dyes are used annually in Japan [2], and 10 to 15% of these dyes are estimated to be discarded into the environment [3]. As colored effluents from major textile and dyestuff industries are recalcitrant to biodegradation, the treatment of these effluents in wastewater treatment systems is mainly based on physical and/or chemical procedures, such as adsorption, concentration, chemical transformation, and incineration. Although these treatment methods are effective, they have several shortcomings, such as high cost, formation of hazardous byproducts, and intensive energy requirements. Therefore, biological degradation methods are receiving attention as better alternatives. Several strains as dye-degrading microorganisms have been reported, such as white-rot fungi *Phanerochaete chrysosporium* [4, 5], *Pleurotus ostreatus*

Makoto Shoda
All rights reserved-© 2024 Bentham Science Publishers

[6], *Coriolus versicolor* [7], and *Streptomyces* spp [4]. The effectiveness of microbial treatment depends on the survival, adaptability, and stable activity of the selected microorganisms in the treatment environment.

The newly isolated strain, Dec 1 decolorized various reactive dyes, including azo and anthraquinone dyes, as shown in Chapter 2. The involvement of several extracellular enzymes, such as Lignin Peroxidase (LiP) and Manganese Peroxidase (MnP), was suggested by the broad decolorization spectrum of this strain. LiP, MnP, Laccase (Lac), and Horseradish Peroxidase (HRP) have also been reported to have the ability to decolorize various dyes [8 - 14]. Among them, MnP from *P. chrysosporium* is a representative enzyme and plays a major decolorizing role in the presence of manganese ions [15, 16]. However, Mn-oxidizing peroxidases isolated from *Bjerkandera adusta* and *Pleurotus eryngii* decolorized with multiple azo dyes, regardless of the presence of manganese ions [17]. Those enzymes were able to oxidize Mn^{2+} to Mn^{3+} at pH 5 and also oxidized aromatic compounds, such as Veratryl Alcohol (VA), a typical substrate of LiP, at pH 3, regardless of the presence of Mn^{2+} [18, 19]. Therefore, these enzymes that expressed both LiP-like and MnP-like characteristics were named MnP–LiP hybrid peroxidase or manganese-independent peroxidase or Versatile Peroxidases (VPs).

The enzymes responsible for the dye-decolorizing activity of Dec 1 were purified. Their characteristics are clarified in Chapter 3. One of them is a new peroxidase, DyP that is a glycoprotein with a molecular mass of 60 kDa, showing a high ability to decolorize anthraquinone dyes [20].

The culture broth of Dec 1 showed the ability to oxidize Veratryl Alcohol (VA), but DyP did not degrade VA [21], suggesting Dec 1 to produce Aryl Alcohol Oxidase (AAO) [22]. Therefore, a veratryl alcohol-oxidizing enzyme was purified from the culture broth of Dec 1, and its enzymatic characteristics and roles in dye decolorization have been characterized *in vivo*, as elucidated in Chapter 3.

Dec 1 showed complete decolorization of anthraquinone dye, Reactive Blue 5 (RB 5) *in vivo*. However, this phenomenon was not observed *in vitro* by DyP alone. DyP changed the color of RB5 from dark blue to a light reddish-brown colored substance composed of an azo complex mixture. Based on this observation, for the complete decolorization of Dec 1, the involvement of other enzymes in addition to DyP has been suggested. Then, a novel Versatile Peroxidase (VP) from Dec 1, named TcVP1, was isolated and characterized. The first complete *in vitro* decolorization of an anthraquinone dye using DyP and TcVP1 [23] is described in Chapter 3.

DyP has two specific characteristics. The first characteristic is its high ability to decolorize anthraquinone dye at around pH 3 and it lacks an important histidine residue that is involved in other fungal peroxidase tertiary structures. Instead, DyP includes aspartic acid and arginine [24], as described in Chapter 4.

The second characteristic is that DyP belongs to a novel DyP-type peroxidase family.

Peroxidases are classified into two types: animal and plant peroxidase superfamilies. The plant peroxidase superfamily is further categorized into three classes according to the origin [25]. Class I peroxidases are prokaryotic, and the representatives are Cytochrome C peroxidase (CCP) and *Escherichia coli* Peroxidase (ECP) [26, 27]. Class II peroxidases are secretory fungal peroxidases and the representatives are *Arthromyces ramosus* Peroxidase (ARP), Lignin Peroxidase (LiP), and Manganese Peroxidase (MnP) [28 - 30]. Class III peroxidases are classical, secretory plant peroxidases, and the representatives are Horseradish Peroxidase (HRP) and Turnip Peroxidase (TP) [31, 32]. According to this classification, DyP belongs to class II. However, the characteristics of DyP are different from those of class II peroxidases. Thus, to clarify the classification of DyP, the gene encoding DyP was cloned from the cDNA library of Dec 1, and the primary structure of DyP was compared with those of other peroxidases. This analysis revealed DyP as a unique peroxidase among previously reported peroxidases. The detail is provided in Chapter 4.

Cloning the cDNA of the *dyp* gene was successful, and the gene encoding DyP was transformed into the host, *Aspergillus oryzae*, under the control of the *amyB* promotor. The massive production of rDyP using recombinant *A. oryzae* is of primary interest in order to use crude rDyP directly for decolorization. This aspect is described in Chapter 4.

A. oryzae was selected as a host because this fungus is recognized as a safe host and it has a high growth rate and can secrete gram-per-liter quantities of heterologous proteins [33].

As the productivity of original peroxidase, DyP, by Dec 1 was extremely low, the enhanced productivity of rDyP was tried. First, the production of rDyP by recombinant *A. oryzae* was enhanced, but it was still not found to be satisfactory [34, 35]. Most common bioprocesses for large-scale production of chemicals use batch culture, which has the advantages of stable nongrowth-associated product formation, maintenance of genetic stability, and a relatively low risk of contamination [36]. Especially when natural substrates are chosen, complex solid particles are involved in the substrates, and continuous cultivation complicates the operation. Thus, the production of the enzyme, rDyP, using different batch cultivation methods and complex substrates is demonstrated in Chapter 4.

As batch culture methods, repeated-batch and fed-batch cultures are the conventional methods of achieving high enzyme production. In repeated-batch culture, the medium is periodically sedimented to draw off the liquid broth, and fresh medium is supplied. This method is effective because the grown microorganisms can be reused and the lag phase associated with at the start of cultivation, and reactor clean-up after cultivation, preculture preparation, and bioreactor sterilization between each batch are unnecessary. For example, in the LiP production by *P. chrysosporium*, repeated batch cultures with 3-4 days cycles were conducted for more than 30 days [37, 38].

In fed-batch culture, fresh medium is supplied intermittently until the culture volume reaches its upper limit. This method is useful in that the by-product formation is minimal, and the energy source is efficiently utilized. Fed-batch culture is worldwide applied when high cell masses are desired, like in the production of baker's yeast [39, 40]. As no fed-batch culture has been reported for the production of fungal enzymes, this method could be more time and cost-effective than repeated-batch culture because it requires less operation time. The comparison of fed-batch and repeated-batch cultures for the production of rDyP is provided in Chapter 4.

Dec 1 requires carbon sources to exhibit decolorization ability for a wide range of colored substances. Molasses is a practically available carbon source because it is widely used in the microbial industry. However, molasses contains highly colored substances, like melanoidins, and the color is deepened by the Maillard reaction after sterilization and remains in Molasses Wastewater (MWW) after its use [41]. Therefore, it is important to decolorize molasses after its use as a carbon source. Several microorganisms have been reported to be effective for the microbial decolorization of MWW [42 - 44], but the decolorizing efficiency is still not satisfactory.

First, molasses was used as a carbon source for the production of DyP, and the effect of molasses on its activity was investigated using a jar fermenter with different sufficient oxygen supply modes [45, 46], as shown in Chapter 3.

Next, the simultaneous decolorization of molasses and anthraquinone dye, RB5, in the molasses medium by Dec 1 was examined to assess the factors involved in the decolorization process [21]. The immobilized cells of Dec 1 on polyurethane foam and suspended Dec 1 cells were compared by repeated-batch culture for long-term simultaneous decolorization of molasses and dye, RB5 [47]. The result is demonstrated in Chapter 6.

The choice of substrate is an important factor in optimizing enzyme productivity. The optimization involves the evaluation of the cost and the availability of nutrients and the reactor system. Repeated-batch culture and fed-batch culture were selected to compare rDyP production using wheat bran powder and rice bran powder as complex substrates, as shown in Chapter 6.

As mentioned above, inexpensive complex media, such as rice bran powder and wheat bran powder, have been found effective for the high production of DyP in the two batch cultures. On the other hand, defined synthetic media are more expensive than complex media, but have more advantages over complex media in that reproducible cultivation performance is possible, process control and monitoring are easier, sterilization conditions are stable, and thus, the scaling-up problem is minimized [48]. The rDyP production in the repeated-batch process of *A. oryzae* pellets using a defined synthetic medium containing maltose as a carbon source increased the number of cycles and avoided the drawback of unstable rDyP activity caused by complex media, as described in Chapter 4. Maltose induced Taka-amylase A production and *A. oryzae* was transformed with the *dyp* gene from Dec 1 using the *amyB* promoter, and thus, high enzyme production became possible [49]. Then, the decolorization of an anthraquinone dye, Remazol Brilliant Blue R, abbreviated as RBBR, using a crude enzyme solution was conducted, as described in Chapter 4.

Among several cultivation methods, liquid cultivation is generally a convenient method for the growth of microorganisms and the isolation of products. The control of cultivation conditions is easy in liquid cultivation because of the uniform culture broth. However, enzyme production hardly occurs in liquid cultivation by the fungus. The liquid cultivation for enzyme production and *in vivo* wastewater treatment is not applicable because large amounts of suspended solids, including fungal flocs, metabolites, and solid particles derived from complex medium, disturb the smooth flow and oxygen transfer. On the other side, Solid-state Fermentation (SSF) or Solid-state Cultivation (SSC) is widely applied in the fungal industry for the production of metabolites from fungi. In Japan, traditional industrial cultivation with *A. oryzae* has been conducted using SSC.

The authors showed that even as a bacterium, *Bacillus subtills* produced lipopeptide antibiotics 10 times higher using SSC compared to liquid culture [50]. One of the advantages of SSC is that fungal surface growth under SSC conditions leads to high enzyme productivity. Thus, protein productivity per unit of dry cell mass in SSC has been reported to be higher than that in liquid culture [51]. The production of β-fructofuranosidase, feruloyl esterase, and lipase from SSC by *Aspergillus niger* was higher than that from liquid culture [52 - 54]. However, some problems remain to be solved in SSC. The extraction and isolation of secreted enzymes in SSC are difficult because of the lack of an aqueous phase and the use of rich suspended solids. In SSC, it is difficult to control incubation conditions, such as temperature, pH, and moisture content, because of heterogeneous solid culture broth.

To resolve these disadvantages of SSC and liquid culture in fungal cultivation, an Air Membrane Surface (AMS) bioreactor was introduced [55, 56]. This reactor system is a hybrid of liquid and solid-state culture. The production of a neutral protease from *A. oryzae* was 10-fold higher than using liquid culture [55], and smooth repeated-batch cultures were successful in accelerating enzyme production because the solid and aqueous phases are separated in AMS [57, 58]. Therefore, the AMS reactor allows easy isolation of soluble secreted proteins in the aqueous phase without contamination of suspended solids. Additionally, this method saves energy. Therefore, Dec 1 cultures using AMS bioreactor have been reported to be focused on the effect of enzyme productivity on *in vivo* dye decolorization and the relation between biofilm formation and water activity. Using the AMS reactor, two new isozymes of rDyP were produced and characterized, as detailed in Chapter 4.

The direct use of enzymes for water treatment is attractive due to the following advantages, such as the compatibility of enzymes, ease of use, simplicity of process control, and wide applicability to different contaminants and pHs [59, 60]. However, despite the potential industrial use of enzymes, their practical application is difficult because enzymes are available in limited amounts, and enzyme stability under unstable environments is low [61]. The *dyp* gene encoding DyP was transformed into *A. oryzae* to produce rDyP, and the enhanced production of rDyP by more than 100-fold compared to the production of Dec 1 was demonstrated. Furthermore, the selection of proper cultivation methods can increase the rDyP production level by nearly half a million times than the original level of DyP, leading to the possibility of using DyP practically, as described in Chapter 4.

When rDyP is applied in industrial processes, some constraints, such as maintenance of its stability, reuse of rDyP, and inactivation by H_2O_2, have to be considered [62]. To overcome these constraints and enhance the catalytic efficiency of rDyP, immobilization of rDyP is a potential method. Several enzyme immobilizations have been attempted. They include adsorption to activated carbon and glass beads, cross-linking by glutaraldehyde, and entrapping by photosensitive resins, but these methods for immobilization of rDyP have not been successful, presumably because of the completely new tertiary structures of rDyP relative to well-known peroxidases, as highlighted in Chapters 3 and 4 [23, 63]. In addition, ordered mesoporous materials have been applied for rDyP because they are new candidates for the immobilization of enzymes due to their uniform and adjustable pore size in the range of 1-10 nm, large surface area, pore volume, mild immobilization conditions, and efficient physical adsorption [64 - 66]. Mesoporous materials, FSM-16 and AlSBA-15, were synthesized using cationic and non-ionic surfactants, respectively, and immobilization of rDyP was attempted on these materials. RBBR was decolorized by immobilized rDyP on these materials using repeated-batch mode, as presented in Chapter 5.

When rDyP is used for dye decolorization, the decolorization activities of rDyP are affected due to the presence of several factors, such as organic acids, pH, temperature, and salts. Among them, inactivation by excess H_2O_2 is the most critical factor [67].

In order to obtain a quantitative evaluation of rDyP inactivation by H_2O_2, the potential turnover capacity of crude rDyP was introduced in batch and fed-batch cultures [62]. The turnover capacity is defined as the unit concentration of reducing substrate removed from the solution per unit concentration of inactivated enzyme [68]. In order to determine the potential of rDyP when the enzyme is subjected to H_2O_2 in different ways, the turnover capacity is useful, as described in Chapter 5. The capacity can compare the ability of rDyP with other dye decolorization methods, such as adsorption, oxidative processes, and microbial treatment.

Another possible application for Dec 1 is the treatment of kraft pulp bleaching effluent that contains colored substances and chlorinated organic compounds, causing environmental pollution because of their poor biodegradability and toxicity [69, 70]. Several methods, such as lime coagulation, rapid land filtration, membrane processes, and polymeric adsorbents, have been adopted to remove these waste products [71]. However, in general, these processes have several problems, such as high energy costs and the generation of hazardous by-products. As microbial treatments, white rot fungi, *P. chrysosporium*, *C. versicolor*, *Pleurotus ostreatus*, and *Streptomyces* spp. have been reported [72 - 75]. Although the microbial methods are energy-saving, some fungi are plant pathogens, have limited substrate specificity, and have a slow growth rate.

On the other hand, non-pathogenic Dec 1 secretes beneficial enzymes and shows a faster growth rate than other white rot fungi. Therefore, Dec 1 was applied to decolorize kraft pulp bleaching effluent to evaluate its effectiveness for the removal of colored substances, as described in Chapter 6.

The multistage bleaching process using chlorine-based chemicals was employed to remove existing lignin in Unbleached Kraft Pulp (UKP) in the paper industry. However, environmental pollutants, such as dioxins, are included in effluent drained from chlorine-based bleaching process [70, 76]. The oxygen bleaching process is a developed process to reduce the consumption of chlorine-based chemicals [77]. However, the removal of colored substances, including lignin, is still unsatisfactory in Oxygen-delignified Bleaching Effluent (OBE) drained from the oxygen bleaching process. As the content of residual lignin of Oxygen-delignified Kraft Pulp (OKP) is half of UKP, biobleaching is applicable. For the biobleaching, *Trametes versicolor* [78], a fungus IZU-154 [79], *Bjerkandera* sp. strain BOS55 [80], and a fungus SKB-1152 [81] showed the possibility of removing the lignin of pulp. Here, a decolorizing non-white rot fungus, Dec 1, was applied to the decolorization of OBE and the bleaching of OKP, as described in Chapter 6.

Dec 1 was initially identified as *Geotrichum candidum.* This fungus was introduced to have biotechnological importance [82] and has been used in mixed culture with other fungi for reducing environmental issues [83].

Since DyP was purified from fungus Dec 1 and proposed as a new DyP-type peroxidase [21, 24], the enzyme DyP was classified as a new family of heme peroxidase. DyP-type peroxidases were then isolated from various microorganisms and characterized. Bacterial genes of P class DyP and a phylogenetic tree with all available sequences of DyP genes offered an overview of their presence in the bacteria kingdom [84]. Ligninolytic bacteria, *Raoultella ornithinolytica* OKOH-1, showed DyP-type peroxidase degradation of congo red and melanin [85]. A salt-tolerant yeast, *Galactomyces geotricum*, degraded red azo dyes using two lignin peroxidase and laccase [86]. A new DyP from thermophilic actinomycete, *Thermomonospora curvata* (*Tc*DyP), was then identified and characterized. *Tc*DyP is highly close to fungal DyPs with the unique catalytic property of A-type DyPs [87]. *Streptomyces albidoflavus* BSII#1 oxidized seventeen phenolic substrates with actinobacterial DyP type peroxidase [88].

We also isolated a new versatile peroxidase VP from Dec 1 named TcVP1 and characterized it, and then, the first complete *in vitro* decolorization of an anthraquinone dye using DyP and TcVP1 was demonstrated [23]. Dye-decolorizing Peroxidase (DyP) from *Auricularia auriculajudae* and Versatile Peroxidase (VP) from *Pleurotus eryngii* oxidized mononitrophenol isomers, and the oxidation site was located at residue in both DyP (Trp377) and VP (Trp164) [89].

DyP produced from *Vibrio cholerae* catalyzed the anthraquinone dyes with acidic optimal pH and identified the decisive factor of the low pH [90]. Fungal DyPs showed efficient lignin biodegradation [91].

On the other hand, immobilization of DyP was found effective in enhancing the decolorization rate and efficiency, as described in detail in Chapters 3 and 4. The agarose-immobilized MnP exhibited a remarkable bioremediation potential as a biocatalyst using a packed bed reactor system [92]. Some books and papers have summarized the latest advancements in colored wastewater treatment, and introduced DyP to have potential application as a biocatalyst in advanced biological processes [93 - 96].

CHAPTER 2

Characteristics of a New Fungus Isolated for Dye Decolorization

Abstract: As a dye-decolorizing microorganism, a fungus, identified as *Geotrichum candidum* Dec 1, was isolated from the soil. Re-identification of this fungus revealed it to be *Thanatephorus cucumeris* Dec 1 abbreviated as Dec 1. A crude extracellular enzyme solution was prepared from the culture of Dec 1, and the enzyme solution showed a broad decolorization spectrum of dyes in the presence of H_2O_2, indicating the existence of multiple enzymes, including peroxidases. Dec 1 decolorized approximately 12 g/l of the dye, Reactive Blue 5 (RB5), without a decline in the decolorizing activity at the inhibitory concentration for most of the microorganisms, indicating the resistant property of Dec 1 to a high concentration of the dye RB5. Dec 1 expressed decolorizing activity in the presence of carbon sources and oxygen.

Keywords: *Geotrichum candidum* Dec 1, *Thanatephorus cucumeris* Dec 1, Peroxidase, Reactive Blue 5, Veratryl Alcohol, Azo dye, Anthraquinone dye.

1. ISOLATION OF A NEW DECOLORIZING FUNGUS

1.1. Methods

1.1.1. Isolation of Dye-decolorizing Microorganisms

For the isolation of dye-decolorizing bacteria and fungi, the following media were used: GPY agar for the isolation of bacteria and PDA media for the isolation of fungi. GPY agar medium consists of glucose, Polypeptone, yeast extract, KH_2PO_4, $MgSO_4.7H_2O$, and agar (pH=7). PDA medium contains potato extract, glucose, and agar (pH=5.6). Each of the following dyes, Reactive blue 114, Acid blue 324, and Dispersive blue 79, were mixed with the media shown above, and microbial colonies that appeared on the media and formed transparent dye zones on the media were purified. Each purified microorganism was used to confirm dye decolorizing ability by the method described below.

Makoto Shoda
All rights reserved-© 2024 Bentham Science Publishers

1.1.2. Structures of the Dyes Used

Table **1** lists the dyes and simplified structures of the dyes used. The simplified structures of Reactive Blue 5 (RB5) are AQ-1[1-amino-4-(3-amino-4-sodiumsulfonoanilino)-2-sodium anthraquinone sulfonate] and AQ-2 (l-amino-4-methylamino-2-sodium anthraquinone sulfonate). The simplified compound of Reactive red 33 is AZ-1[1-hydroxy-6-methylamino-3-sodium naphthalene sulfonate-2-azo-(4'methoxy-2'sodium benzene sulfonate)]. Fig. (**1**) shows the structures of some dyes and the model compounds used. The simplified structure of AQ-2' is 1,4-diamino-2-sodium anthraquinone sulfonate.

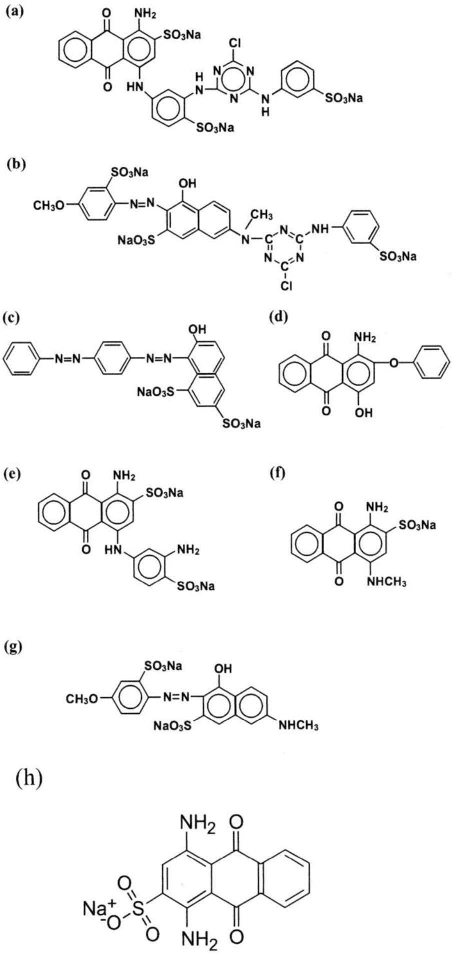

Fig. (1). Structure of representative dyes (**a-d**) and the model compounds of RB5 and Reactive red 33 (**e-h**) used. (**a**) Reactive blue 5; (**b**) Reactive red 33; (**c**) Acid red 73; (**d**) Dispersive red 60; (**e**) AQ-1; (**f**) AQ-2; (**g**) AZ-I; (**h**) AQ-2' [1].

Table 1. Degree of decolorization of various dyes by *T. cucumeris* Dec 1 on PDA medium after 4 day incubation at 28°C.

Color Index	Chromophore	Initial Concentration (ppm)	Decolorized Diameter (cm)[a]	Relative Value[b]
Reactive blue 5	Anthraquinone	100	8.5	1
Reactive blue 19	Anthraquinone	100	8	0.94
Reactive blue 114	Anthraquinone	100	7.5	0.88
Reactive blue 182	Azo	100	8	0.94
Reactive black 5	Azo	100	8	0.94
Reactive red 33	Azo	100	7	0.82
Reactive red 120	Azo	100	5.5	0.65
Reactive red 123	Azo	100	6.5	0.77
Reactive red 187	Azo	100	4	0.47
Reactive red 202	Azo	100	6.5	0.77
Reactive red 225	Azo	100	6	0.71
Reactive orange 13	Azo	100	3.5	0.41
Reactive orange 30	Azo	100	4	0.47
Reactive violet 23	Azo	100	7.5	0.88
Reactive yellow 2	Azo	100	3.5	0.41
Acid red 73	Azo	100	6.5	0.77
Acid blue 324	Azo	100	7	0.82
Disperse red 60	Anthraquinone	100	3	0.35
Disperse blue 79	Azo	100	n.d.	n.d.
AQ-1[c]	Anthraquinone	100	7.5	0.88
AQ-2[c]	Anthraquinone	100	8.5	1
AZ-1[d]	Azo	100	6	0.71

a Decolorized diameter after 4 day incubation.
b Reactive value of diamater to diameter of Reactive blue 5.
c Model chemicals of Reactive blue 5.
d Model chemicals of Reactive red 33.
n.d.Not detected.

1.1.3. Decolorization Using the Purified Dec 1 on PDA Medium

Dyes were mixed with PDA medium and the isolated Dec 1 mycelia were placed on the center of the PDA plate containing each dye; the plates were incubated until their transparent surface appeared.

1.1.4. Decolorization with the Purified Dec 1 in Liquid Medium

A spore suspension of about 10^7 colony-forming units (cfu/ml) was prepared by suspending Dec 1 mycelia obtained from PDA medium in sterile distilled water. This spore suspension and each dye solution were put in Potato Dextrose (PD) liquid medium and cultivated at 30°C.

Periodically, the dye concentration of each culture was monitored by observing the absorbance of each dye. Table **2** shows the maximum absorbance wavelength of the dyes used. The decolorizing rate of each dye was determined by measuring the absorbance difference between the initial and sampling times.

Table 2. Decolorizing rates of dyes determined by culture broth of Dec 1 grown in liquid PD medium and by extracellular crude enzyme solution prepared from the culture broth of Dec 1 in liquid PD medium [1].

Color Index	Maximum Absorbance Wavelength (nm)	Concentration Added (ppm)	Decolorizing Rate (ppm/h)[a]	Decolorizing Rate (ppm/h)[b]
Reactive blue 5	600	200[c] (100)[d]	95	2442
Reactive blue 19	590	150 (70)	42	1740
Reactive blue 114	620	200 (100)	71	702
Reactive blue 182	610	300 (120)	181	3018
Reactive black 5	598	60 (30)	11	20.4
Reactive red 33	500	100 (50)	22	31.4

a: Decolorizing rate by the culture broth of Dec 1.
b: Decolorizing rate determined by extracellular crude enzyme solution (0.5 µg protein/ml).
c: Initial dye concentration used for measuring decolorizing rate by the culture broth of Dec 1.
d: Initial dye concentration used for measuring decolorizing rate by the crude enzyme solution (0.5 µg protein/ml).

1.1.5. Extracellular or Intracellular Crude Enzyme Solutions Prepared from Dec 1

The cells of Dec 1 and the dye, RB5, were mixed in a PD liquid medium, and the culture was incubated for 3 days. The supernatant obtained from the culture broth was precipitated to obtain crude proteins by adding ammonium sulfate powders. The protein solution was dialyzed and the dialyzed solution was considered as an extracellular crude enzyme solution. On the other hand, the Dec 1 cell mass was disrupted using quartz sand and the disrupted cells were centrifuged, and an intracellular crude enzyme solution was obtained from the supernatant.

1.2. Results

1.2.1. Decolorization by the Isolated Fungus, Dec 1, on Solid Medium

Among isolated fungi and bacteria, one fungus showed both high dyedegradation ability and a broad dye-degrading spectrum. This fungus was identified as *Geotrichum candidum* Dec 1. Later, reidentification verified this fungus as *Thanatephorus cucumeris* Dec 1. Therefore, this fungus was named as Dec 1 in all experiments. Table **1** shows the result of the decolorizing *in vitro* test of Dec 1 using 19 kinds of anthraquinone and azo dyes together with the three model compounds of RB5. Fig. (**2**) shows photographs of the decolorized three dyes on PDA medium by Dec 1.

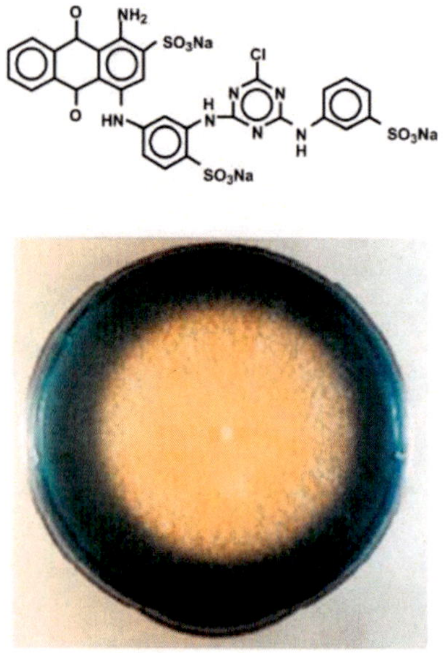

C.I. Reactive blue 5
Chromophore:anthraquinone

(Fig. 2) contd.....

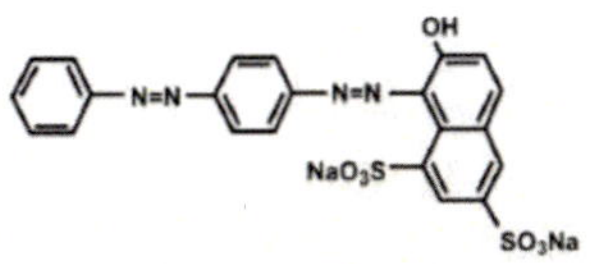

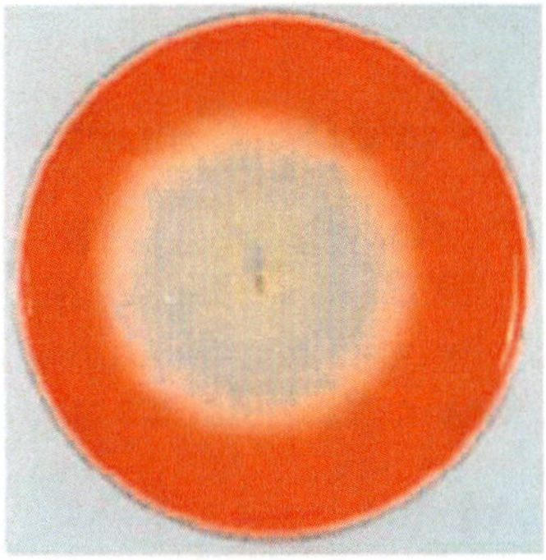

C.I. Acid red 73
Chromophore:azo

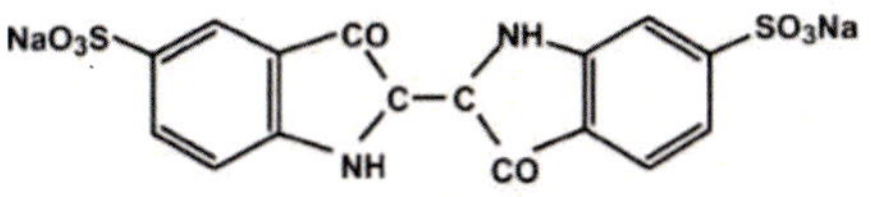

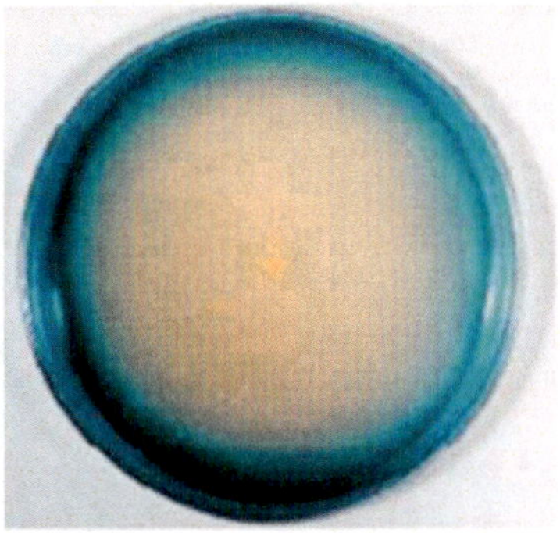

Acid blue 74
Chromophore:indigo

Fig. (2). Decolorizing activity of Dec 1 *in vitro* on three dyes containing anthraquinone, azo and indigo as chromophores (photos have been taken by the author).

Table **1** also shows decolorization of the simplified model compounds of RB5, AQ-1 and AQ-2, and the simplified model compound of Reactive red 33, AZ-1. This indicates that Dec 1 attacks the chromophores of dyes directly. Table **1** also shows the degree of decolorization as relative values. The different decolorizing degrees reflected that the effectiveness of decolorization by Dec 1 was different depending on the structure and complexity of each dye.

1.2.2. Decolorization by the Isolated Dec 1 in Liquid Medium

Table **2** shows the decolorizing rates of the dyes by the liquid culture medium of Dec 1 and also by crude extracellular enzyme solutions prepared from the culture broth of Dec 1. The most decolorizing rate measured with crude enzyme solutions was 40-fold higher than that measured with the liquid culture broth of Dec 1.

Fig. (**3**) demonstrates changes in the growth of Dec 1 and the pH of liquid culture broth. The Dec 1 grew for 3 days and then its growth declined because of autolytic degradation of the cells. The pH value was stable around 5 for 2 weeks.

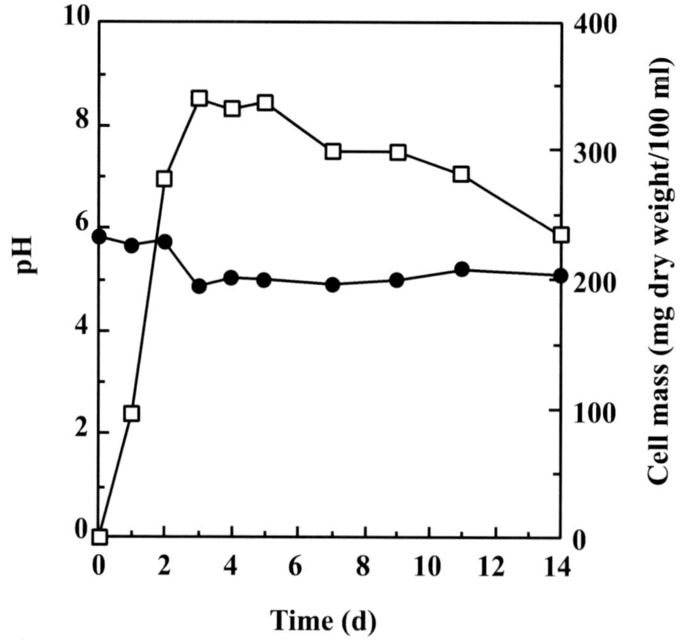

Fig. (3). Change in pH (●) and growth of Dec 1(□) in liquid PD medium at 30°C [1].

Fig. (**4**) shows changes in the decolorizing rates of RB5 and the glucose concentrations in the medium. After the logarithmic growth phase, the decolorizing activity appeared quickly, and the activity maintained until 14 days. The activity deteriorated significantly due to lowered glucose concentration. By supplementing 10 g/1 glucose on day 19 as indicated by a black arrow, the decolorizing activity was quickly recovered. This means that carbon sources used for growth and energy production are essential to maintain the decolorizing activity.

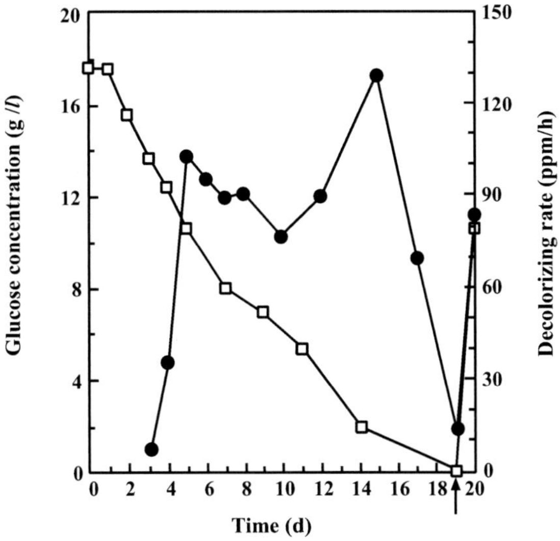

Fig. (4). Change in decolorizing rate of RB 5 dye (•) and glucose concentration (□) when Dec 1 was grown in liquid PD medium. The black arrow indicates the time when 10 g/1 of glucose was added [1].

In separate experiments, Dec 1 utilized acetate, ethanol, and acetone as carbon and energy sources. These materials can be considered as candidates for carbon sources in practical decolorization treatment besides sugars.

Fig. (**5**) shows the change in the decolorization rates of RB5 under different initial RB5 concentrations. In the initial range of 400-1,600 ppm, the decolorizing rate steadily increased. However, at 3,200 ppm, slower decolorization of the dye was observed presumably due to temporal inhibition at the high dye concentration of the cell activity, but the decolorization rate resumed after 4 h.

When 3,000 ppm of RB5 was added repeatedly, a total of 12 g/l of RB5 was decolorized up to the fourth addition of the dye before a significant decline in activity, as shown in Fig. (**6**). When initial 12 g/l of the dye was added to the medium, Dec 1 decolorized almost all of the dye within 2 days. This result suggests Dec 1 to be resistant to degradation at a high dye concentration.

The optimal temperature for the decolorizing activity was found to be 30°C.

When N_2 gas was supplied to the culture, no decolorization activity was observed, and when O_2 gas was supplied to the culture, decolorization was observed. This suggests that aerobic condition is essential for Dec 1 to express dye-degradation activity.

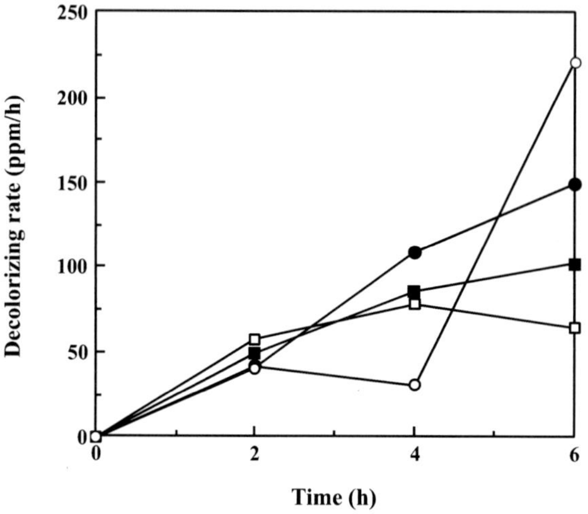

Fig. (5). Change in decolorizing rates when various initial dye concentrations of RB5 were added to liquid PD medium. □, 400 ppm; ■,800 ppm; ●,1,600 ppm; ○, 3,200 ppm [1].

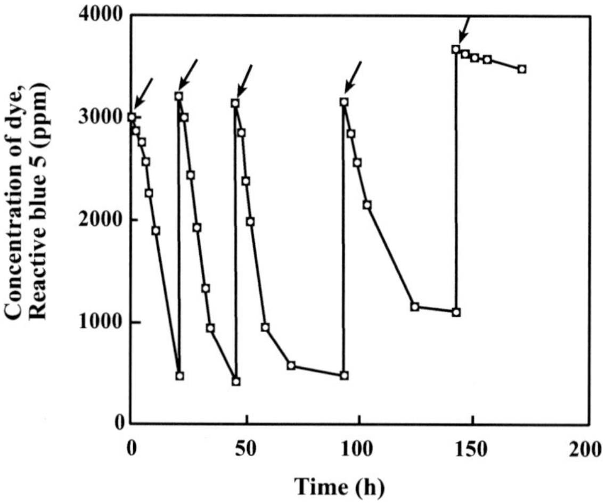

Fig. (6). Repeated degradation of RB5 when 3,000 ppm was added each time (shown by black arrows) [1].

1.2.3. Decolorization by Crude Enzyme Solution

The intracellular crude enzyme solution prepared from the cells of Dec 1 was mixed with H_2O_2 and RB5, but no dye-decolorization activity occurred. Fig. (**7**) shows the decolorizing activity of RB5 by the extracellular crude enzyme solution when a mixture of H_2O_2 and RB5 was supplied. No decolorization activity was observed without H_2O_2, or when crude enzyme solution was denatured at 85°C. The data confirmed the existence of peroxidases in the extracellular crude enzyme solution. Table **2** shows the decolorizing rate of culture broth of Dec 1 and extracellular crude enzyme solution.

Two peaks of decolorizing rates were seen during the cultivation period, as shown in Fig. (**4**). This suggests the possibility of the production of different kinds of peroxidases or the involvement of enzymes other than peroxidases. As Dec 1 showed the ability to degrade veratryl alcohol, which is a substrate for the activity of lignin peroxidase, the details of other enzymes produced by Dec 1 were analyzed and are described in the next chapter.

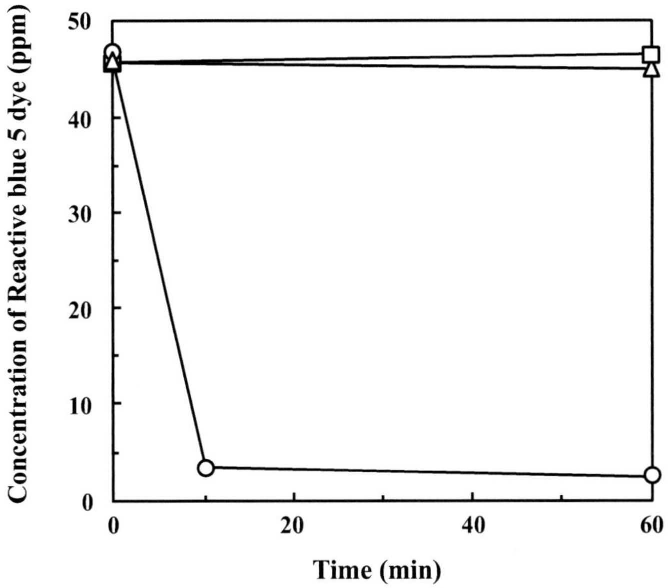

Fig. (7). Decolorization of RB5 by extracellular crude enzyme solution in the presence of H_2O_2 (○) and in the absence of H_2O_2 (□), and by heat-denatured enzyme solution in the presence of H_2O_2 (Δ) [1].

1.3. Discussion

Several dye-decolorizing microorganisms have been reported. They are *P. ostreatus* [6], *C. versicolor* [7], *Streptomycetes* [97], *P. chrysosporium* [5, 98, 99], *Streptomyces* sp., and *P. chrysosporium* [4, 100]. As some fungi oxidized lignin-related compounds [101] or herbicides [102], the existence of peroxidase-like enzymes was suggested.

Decolorization of various dyes by a newly isolated *T. cucumeris* Dec 1 was observed both in solid and liquid cultures, as shown in Tables **1** and **2**. Extracellular peroxidase-type enzymes, including DyP, are involved in a broad decolorizing spectrum by Dec 1. The details of DyP and other enzymes are described in the next chapter.

The advantages of Dec 1 over other dye-decolorizing microorganisms are as follows: Dec 1 is non-pathogenic, versatile, and has resistant properties to high concentrations of dyes because 12g/l of RB5 was decolorized without significant decline in the decolorizing rate. Generally, *T. cucumeris* is widely found in the soil and water with a relatively higher growth rate than other fungi at a pH range of 4-7 and a temperature range of 20-30°C. Furthermore, Dec 1 can utilize less expensive carbon sources as well as sugars. These data exhibit the potentiality of Dec 1 to treat colored materials in a reactor system or during bioremediation in soil under properly controlled conditions.

In most conventional wastewater treatments, the use of living cells is dominant. However, enzymatic treatment can be attractive as an alternative compared to the conventional methods. Lignin peroxidases and horse radish peroxidases have been tried in basic research [103, 104]. When the decolorizing activity between the culture broth of Dec 1 and the extracellular crude enzyme solution was compared, the degradation rate of the enzyme solution was significantly higher for some of the dyes, as shown in Table **2**. This suggests that the use of enzymes is effective compared to the use of cells of Dec 1. As the massive production of DyP is possible as shown in chapter 4, based on basic analysis of the enzyme and proper reactor system, this may lead to the direct application of enzymes to the decolorization of several colored materials. In the next chapter, the possibility of massive production of an extracellular peroxidase, DyP, is presented.

The decolorizing rate varied among dyes used, as shown in Tables **1** and **2**. This indicates that the effectiveness of decolorization depends on the structural difference of each dye. This also implies that the use of multiple enzymes is important for the efficient decolorization of colored materials.

As two peaks of activity during the cultivation period of Dec 1 were observed, as shown in Fig. (**4**), the production of different kinds of peroxidases is suggested. The purification and characterization of these enzymes in relation to dye-decolorization ability are described in the next chapter.

Dec 1 degraded veratryl alcohol, exhibiting the properties of lignin peroxidase. Lignin Peroxidase (LiP) and Horseradish Peroxidase (HRP) were reported to degrade nonspecifically many aromatic and substituted aromatic rings [99, 103, 104]. As lignin-degrading ability is related to dye-degradation ability [101], the actual degradation of lignin by Dec 1 was shown in Chapter 6 and the analysis of related enzymes are presented in Chapter 6.

Some experiments showed the rate of decolorization to be slower in the nitrogen-sufficient medium than in the nitrogen-insufficient medium [1]. As Dec 1 also showed dye-decolorization ability in the nitrogen-limited medium, nitrogen limitation can be considered another factor for Dec 1 to express genes responsible for the production of peroxidase. A similar result was reported in the lignin-degrading activity of *P. chrysosporium* [5, 99]. No need for excessive nutrients may result in being advantageous for the growth and the application of the cells.

CONCLUSION

As a dye-decolorizing fungus, *Thanatephorus cucumeris* Dec 1 was isolated and its characteristics were clarified. Its broad dye-degrading ability and tolerance to high concentrations of the dye were confirmed. As Dec 1 decolorized the simplified structures of model compounds of the dyes, Dec 1 can be considered to attack the chromophores directly. Dec 1 demands the use of oxygen and H_2O_2 in dye decolorization and extracellular enzyme solution also demands the use of H_2O_2. This suggests that peroxidases are involved in dye degradation by Dec 1.

CHAPTER 3

Characterization of Multi-Enzymes Produced by the Fungus, Dec 1, Responsible for Dye Decolorization

Abstract: A peroxidase, DyP, produced by the fungus Dec 1 having the ability to decolorize dyes was purified. DyP contains 17% sugar comprising GlcNAc and Man. The molecular mass of DyP was estimated to be 60 kDa and the isoelectric point (pl) was determined to be 3.8. DyP degraded various dyes, and also phenolic compounds of 2,6-dimethoxyphenol and guaiacol. However, veratryl alcohol, abbreviated VA, a substrate of lignin peroxidase, was not degraded by DyP.

Molasses was utilized by Dec 1 to produce DyP as carbon and energy source. Within 10 g/l of molasses concentration, decolorization activity of Dec 1 toward RB5 gradually increased. However, at more than 20 g/l of molasses concentration, the inhibitory effect on the decolorizin g activity of Dec 1 was observed. As the activity of purified DyP was inhibited by 10 g/l of molasses concentration, this indicates that molasses has a stimulative effect on the decolorization activity of Dec 1, but has an inhibitory effect on the DyP activity.

Dec 1 also produced Aryl Alcohol Oxidase (AAO). The role of AAO in the Dec 1 decolorization process of dyes has been observed to be as follows: the first role was that H_2O_2 was produced by AAO oxidation of Veratryl Alcohol (VA) to veratraldehyde and then utilized by the peroxidase DyP. In the cultivation of Dec l, the presence of H_2O_2 and veratraldehyde has been detected. The second role was that the polymerization of products produced by DyP oxidation of a simplified RB5, AQ-2', was prevented by AAO. This was confirmed by the result that the molecular weight of the products was reduced in the mixed decolorization of DyP and AAO.

A new versatile peroxidase named TcVP1 was purified from the Dec 1 culture. Purified TcVP1 behaved as Manganese Peroxidase (MnP) at pH 5, and the enzyme functioned as a Lignin Peroxidase (LiP) at pH 3. As TcVP1 decolorized preferentially azo dyes, co-application of TcVP1 and DyP conducted complete decolorization of anthraquinone dye, RB5, *in vitro* to colorless products.

Keywords: Peroxidase, DyP, Anthraquinone dye, Aryl alcohol oxidase, *Thanatephorus cucumeris* Dec 1, Versatile oxidase, TcVP1, Simplified RB5, AQ-2', Veratryl alcohol, VA, Veratraldehyde, Hydrogen peroxide, Molasses, Manganese peroxidase, MnP, Lignin peroxidase, LiP.

Makoto Shoda
All rights reserved-© 2024 Bentham Science Publishers

1. NEW PEROXIDASE, DYP

1.1. DyP Purification

The purification steps for a new peroxidase, DyP, are shown in Table **3**. The total DyP activity in the supernatant derived from Dec 1 cultivation increased from 24U to 420 U after ultrafiltration with a YM 40 membrane treatment. Fig. (**8**) shows the main peak of DyP activity after the Butyl Toyopearl chromatography treatment.

Table 3. Purification steps for peroxidase, DyP, from cell-free culture of Dec 1 [20].

Purification Step	Protein (mg)	Total Activity (U)	Sp Act (U/mg)	Activity Yield (%)	Purification (fold)
Supernatant	88	24[a]	0.28[a]		1
Ultrafiltration (YM 10)	28	4.2×10^2	15	100[a]	55
Super Q chromatography	7.8	3.0×10^2	34	71	140
Butyl Toyopearl chromatography	1.5	85	57	20	210

Sp Act: specific activity.
a: The activity yield was calculated based on the value after the YM 10 treatment.

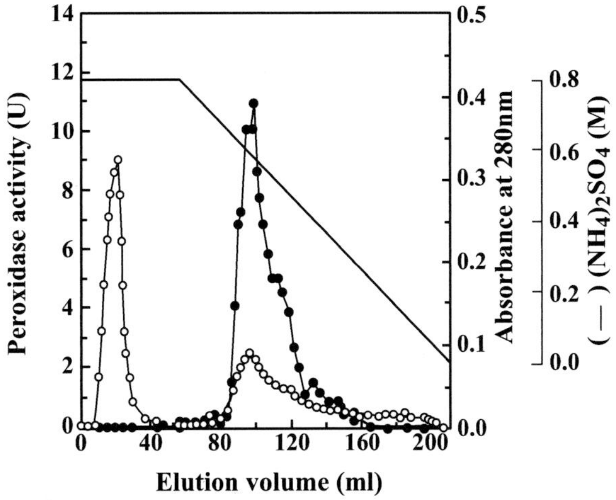

Fig. (8). Purification of peroxidase, DyP, produced by Dec 1 during Butyl Toyopearl chromatography. A solid line is a linear gradient of 0.8 to 0 M $(NH_4)_2SO_4$ in citrate buffer.
Symbols: •, DyP activity; ○, absorbance at 280 nm [20].

Fractions 86 to 99, corresponding to the left half of the main peak, were called purified DyP having a specific activity of 57 U/mg of protein, as shown in Table **3**. SDS-polyacrylamide gel electrophoresis determined the molecular mass of DyP as 60 kDa. That value was significantly larger than those reported previously [105

- 107], ranging from 40 to 44 kDa. The Isoelectric point (*p*I) of DyP was determined as 3.8 by isoelectric point electrophoresis.

1.2. Spectral Characteristics of DyP

Fig. (**9**) shows the spectral characteristics of purified DyP. A Soret band appeared at 406 nm, representing the absorption peak of peroxidase, and the molar extinction coefficient of the band was almost the same as those of other peroxidases previously reported [108 - 110]. The A_{406}/A_{280} RZ value, which represents the purity and spectral characteristics of DyP, was 1.6 at pH 5. After H_2O_2 oxidation of the purified DyP, the peaks at 406 and 510 nm shifted to 400 nm and 530 nm, respectively. Then, $Na_2S_2O_4$ was added to reduce DyP that was oxidized by H_2O_2, and a peak at 556 nm appeared, as can be seen in the inset figure in Fig. (**9**). This confirms that DyP has a heme-pyridine complex, meaning that its prosthetic group is protoheme. DyP holds a single heme because the estimated heme content is 0.6 per mole of DyP.

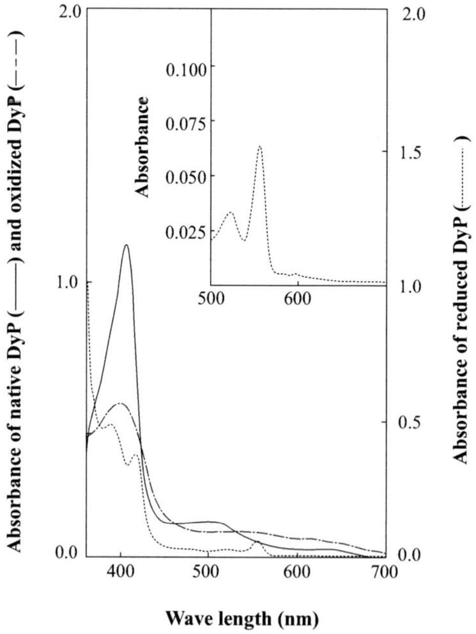

Fig. (9). Spectral characteristics of the following three DyPs: purified DyP, DyP oxidized by H_2O_2, and DyP reduced by $Na_2S_2O_4$. The inset shows reduced DyP in the form of a heme-pyridine complex observed at 556 nm [20].

1.3. Substrate Specificity of DyP

Table **4** shows the decolorization by DyP of four azo dyes, three anthraquinone dyes, and three simplified model compounds of RB5, AQ-1, AQ-2, and AQ-2'.

The structures of AQ-1, AQ-2, and AQ-2' are shown in Chapter 2. DyP demonstrated higher decolorizing activity for anthraquinone dyes than azo dyes. DyP degraded, without the addition of Mn^{2+} phenolic compounds, 2,6-dimethoxyphenol and guaiacol, a substrate of Manganese Peroxidase (MnP). However, veratryl alcohol, a substrate of Lignin Peroxidase (LiP) was not degraded by DyP. As the DyP activity for 2,6-dimethoxyphenol and guaiacol was not enhanced by the addition of Mn^{2+}, it is obvious that DyP has different substrate specificity compared to LiP, MnP, and HRP. Therefore, DyP, as a novel decolorizing peroxidase, is discussed from genetic and structural points in the next chapter.

DyP degraded only 7 dyes of the 18 dyes that Dec 1 decolorized, as shown in Chapter 2. This indicates the possibility that Dec 1 produces other enzymes from DyP that may be associated with a broad decolorizing spectrum. The details of other enzymes are described in the later sections.

Table 4. Decolorization by purified DyP on several dyes and model compounds of RB5 [20].

Color Index of Dyes	Chromophore	ε at λ_{max} [a] (liters/mM/cm)	Optimal pH	Initial Concentration (μM)	Decolorizing Rate (μM/min)
RB5	AQ[b]	7.3	3.2	119	24
RB19	AQ	10	3.2	116	22
RB114	AQ	8.4	4.0	151	12
AQ-1	AQ	9.0	3.2	113	10
AQ-2	AQ	8.3	3.0	141	55
AQ-2'[d]	AQ	7.3	3.2	148	20
Reactive black 5	AZ[c]	37	3.2	33	1.1×10^{-1}
Reactive red33	AZ	23	3.2	60	4.7×10^{-1}
Reactive yellow 2	AZ	8.1	3.2	125	6.2×10^{-1}
RB182	AZ	7.3	4.0	123	21

a: Molar extinction coefficient at maximum absorption wavelength (λmax) of each dye. **b**: AQ, anthraquinone. **c**:AZ, azo. **d**: Decolorization test of AQ-2' by Dec 1 was not conducted in the previous chapter 2.

1.4. Sugar Analysis

In sugar content analysis, DyP was found to contain N-acetylglucosamine (GlcNAc) at 38 mol per mol of DyP and Mannose (Man) at 26 mol per mol of DyP, which indicated DyP to be a 17% (wt/wt) sugar containing glycoprotein.

This value was similar to Horseradish Peroxidase (HRP) containing 16 to 18% sugar [31, 108, 109] and Lignin Peroxidase (LiP) containing 17 to 39% sugar [111]. LiP had both N-linked sugar chains and O-linked monosaccharides [112], while HRP retained only asparagine N-linked types [31]. The detailed structure of DyP was clarified by cloning *dyp*, as presented in Chapter 4.

1.5. Effect of Temperature on DyP Activity

Fig. (**10**) shows the effect of temperatures on DyP activity. The optimal temperature was found to be 30°C, and in the range of 15 to 35°C, the activity of DyP was stable, as shown in Fig. (**10a**). A comparison of the thermostability of DyP and HRP activities is shown in Fig. (**10b**). When DyP was treated at 40 and 50°C for 11 h, DyP activity at 30°C did not change. When DyP and HRP were treated at 60°C for 3h, their initial activities were lost by 5% for DyP and 90% for HRP. This suggests DyP to be more thermostable compared to HRP.

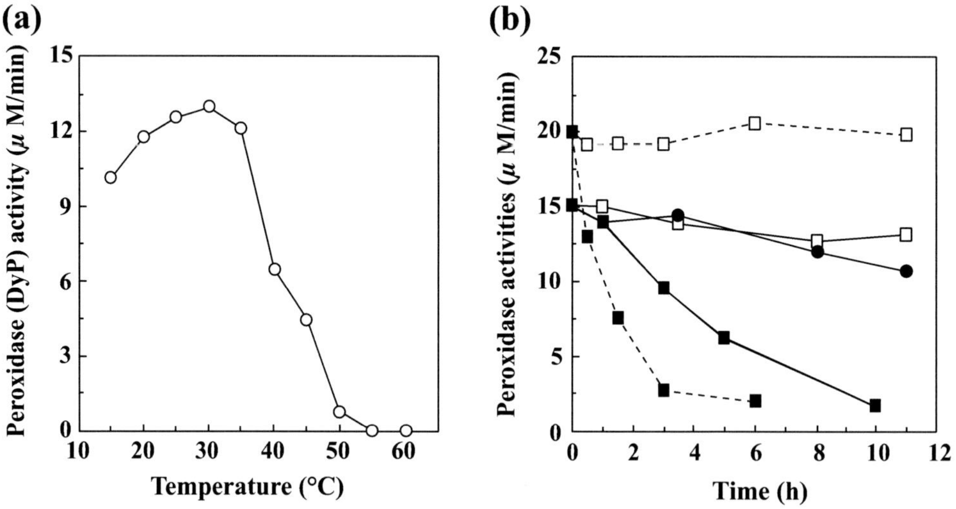

Fig. (10). (**a**) DyP activities at different temperatures. (**b**) Thermostability of DyP and HRP at 30°C and that of DyP after treatment at 40°C (□—□), DyP at 50°C (●—●), DyP at 60°C (■ —■), HRP at 40°C (□ - - - - -□), and HRP at 60°C(■- - - - -■) [20].

1.6. H_2O_2 Inhibition of DyP Activity

Fig. (**11**) shows the inhibitory effects of H_2O_2 on DyP activity. The dye, RB5, and its simplified model compound, AQ-2', were used as substrates for DyP. The inhibition towards RB5 was observed at 0.2 mM H_2O_2 and 0.6 nM DyP. The inhibition toward AQ-2' was seen at 0.1 mM H_2O_2. This indicates that the inhibition degree by H_2O_2 of DyP activity was similar in the presence of different substrates of RB5 and AQ-2'.

The inhibitory effects of H_2O_2 on the activity of HRP or LiP are reported as follows: at first, two electrons from H_2O_2 oxidized these enzymes to compound I, and then, one electron derived from a substrate changed compound I to compound II. Another electron from a substrate reduced compound II to a resting enzyme. However, H_2O_2 in excess forced compound II to convert to compound III, *i.e.*, an inactivated state, and this led to a decrease in these enzymes' activities [113 - 115]. Fig. (**9**) suggests that in DyP, conversion to compound I was reflected by the peak shifts at 406 nm to 410 nm and 510 nm to 530 nm. This has been found similar to the result of LiP [116, 117].

1.7. Comparison of Decolorization Rates of DyP and HRP

The decolorization rates of DyP and HRP toward RB5 were compared by applying the pseudo-first-order kinetics. Table **5** shows the results. $Km_{(RB5)}$ and $Km_{(H2O2)}$ values for DyP were slightly lower than those for HRP. The Kcat(obs) value for DyP was nearly 2 times higher than that for HRP.

When DyP and HRP decolorized RB5 and the resulting solution was scanned, the spectral patterns between them were different, reflecting that the two peroxidases produced different by-products from RB5.

Although the Km values of DyP and HRP toward RB5 were almost the same, the Km value of DyP toward H_2O_2 was lower than that of HRP. This indicates a higher affinity of DyP to H_2O_2 than that of HRP. Kcat(obs) of DyP also showed high decolorization activity of DyP for RB5 due to its higher redox potential.

1.8. DyP Activity Analyzed by Second-order Kinetics

DyP reaction rates toward the two substrates, AQ-2' and H_2O_2, were assessed by second-order kinetics. Lineweaver-Burk plot between AQ-2' concentration and DyP activity provided the apparent value of Km(AQ-2'). Then, the values of Km (H_2O_2) and apparent Kcat were determined from the reciprocal plot of Vmax(app) and H_2O_2 concentration. Table **6** shows the second-order kinetic constants assessed from the above analysis. Kcat/Km(AO-2') and Kcat/Km(H_2O_2) for the substrates reflected the similar reaction rates against the two substrates. As the apparent Kcat value, presented in Table **6**, was similar to that of the Kcat(obs) for RB5, shown in Table **5**, DyP exhibited a similar reaction rate for a simplified substrate, AQ-2'. AQ-2' can be used to compare the mechanism of DyP degradation toward dyes with those of the other peroxidases [8, 117 - 119].

As the values of Kcat of LiP for VA and those of MnP for Mn(II) were reported in the range from 30 to 160 s^{-1} [9, 107, 111], it is apparent that reaction rates of DyP on RB5 or AQ-2 were higher than those reported.

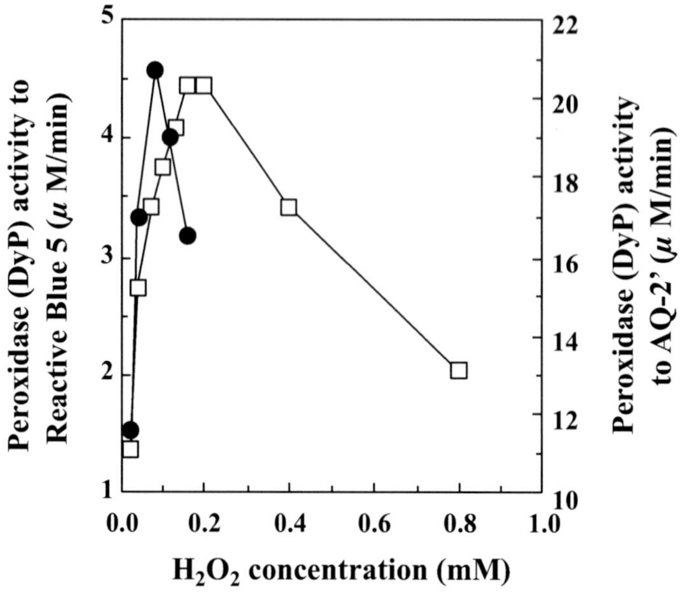

Fig. (11). H_2O_2 inhibition of DyP activity toward RB5 and its simplified model compound, AQ-2'. Symbols: □, the activity of DyP toward RB5; •, the activity of DyP toward AQ-2' [20].

Table 5. Pseudo-first-oder kinetic parameters for activities of DyP and HRP [20].

Parameter	Unit(S)	Value	
		DyP	*HRP*
Km(RB5)	μM	54	58
Km(H_2O_2)	μM	26	36
Kcat(obs)	s^{-1}	260	140
Kcat/Km(RB5)	$M^{-1}s^{-1}$	$4.8x10^6$	$2.4x10^6$
Kcat/Km(H_2O_2)	$M^{-1}s^{-1}$	$1.0x10^6$	$3.8x10^6$

Table 6. Second-order kinetic constants of DyP for H_2O_2 and AQ-2' as substrates [20].

Parameter	Unit(S)	Value
Km(AQ-2')	μM	**84**
Km(H_2O_2)	μM	**36**
Kcat	s^{-1}	**270**
Kcat/Km(AQ-2')	$M^{-1}s^{-1}$	**$3.2x10^6$**
Kcat/Km(H_2O_2)	$M^{-1}s^{-1}$	**$7.6x10^6$**

1.9. DyP Production

Decolorizationrate was maximum at day 6, as shown in Fig. (**4**) in the previous Chapter 2, reflecting maximum production of DyP at day 6. DyP activity was maintained while sugars existed; thus, DyP can be considered a constitutive enzyme. Compared to the production of other fungal enzymes, the production of DyP by Dec 1 was found efficient and simple, because DyP was constitutively produced both under low-shear stress conditions during shaking cultivation [1] or under high-stress conditions using a stirred tank reactor [46] without a decline in the activity. On the other hand, LiP production by *P. chrysosporium* required inducers, such as Veratryl Alcohol (VA) and veratryl acid, under limited nitrogen and aerobic conditions [99, 120]. Low shear stress [34, 121] was also found essential to obtain higher activity of the enzyme-producing fungi.

1.10. Effect of Molasses on DyP Production

The productivity of DyP by Dec 1 and the activity of purified DyP were investigated under different initial molasses concentrations [45].

1.10.1. Methods

1.10.1.1. Cultivation of Dec 1

Dec 1 and spore preparation as inoculum have been characterized in Chapter 2 [1]. The molasses used contained (w/w) 33% sucrose, 6.5% glucose, 7.5% fructose, and 0.57 g/kg nitrogen. The basal medium consisted of glucose, ammonium tartrate, KH_2PO_4, and $MgSO_4.7H_2O$. To make the results clear, the initial glucose concentration was increased to 35 g/l in all experiments.

1.10.1.2. Determination of Decolorization Degree of Culture Broth

The decolorization degree of the culture broth (%) was determined by the absorbance difference of the culture broth at 475 nm between the initial value and the value at each sampling time divided by the initial value [44, 122].

1.10.1.3. Decolorization Activity of DyP Toward RB5 Dye

As dye decolorization activity in the molasses medium was difficult to determine because of the black-colored molasses, DyP activity that showed the ability to decolorize RB5 was used. The supernatant of the culture, RB5, and H_2O_2 were mixed, and the decolorization rate of RB5 was determined using a previously described method [1]. The decolorization rate was expressed as mg/l /min.

1.10.1.4. DyP Purification

DyP in the culture broth of Dec 1 was purified by the method described in a previous section [20].

1.10.1.5. Effect of Molasses Concentration on Decolorization Rate of RB5 by DyP

Purified DyP was added to the different concentrations of molasses solutions and the decolorizing activity toward RB5 by DyP was measured by the same method as that mentioned above.

1.10.1.6. Fractionation of Molasses using Ultrafiltration

The 40 or 4 g/l molasses solutions were fractionated by Centriprep 10 membrane ultrafiltration into the fractions based on molecular weight with a molecular mass of more than 10 kDa and that of less than 10 kDa.

1.10.1.7. Gel Chromatography Fractionation of Molasses

Culture broth containing molasses solutions was fractionated by gel chromatography both before and after Dec 1 cultivation.

1.10.1.8. The Dilution Effect of Molasses

The cell-free culture broth after Dec 1 was cultivated in molasses medium was diluted at different levels and the RB5-decolorization activity of each diluted molasses medium was determined by the method described above.

1.10.2. Results

Fig. (**12**) shows a change in the production of DyP by Dec 1 under different initial molasses concentrations. The dye-decolorization activity increased when molasses concentration increased and the maximum activity reached 570 mg/l/min at 40g/l glucose concentration. As the initial glucose concentration was fortified to 35 g/l, the increase in decolorization activity was due to other components than glucose, such as fructose or nitrogenous compounds in molasses [46].

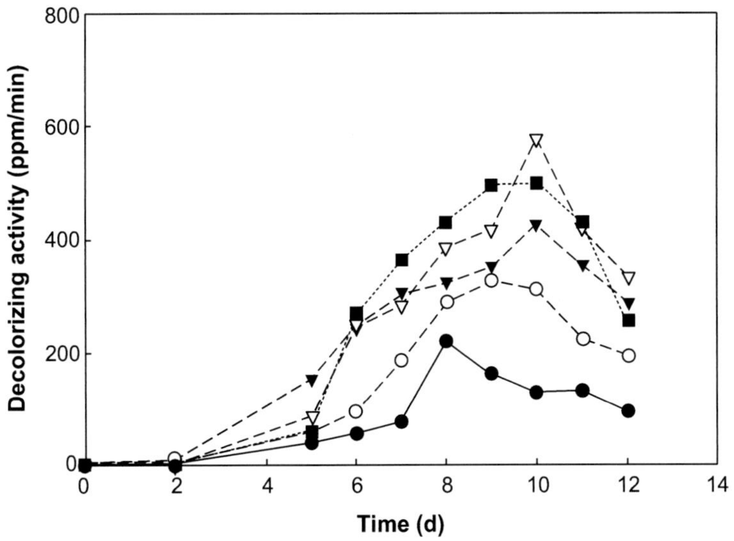

Fig. (12). Change in the production of DyP under different initial molasses concentrations at 35 g/l glucose concentration. Symbols: •,10g / 1 molasses; o, 20 g/l;▼,30 g / l ; ▽,40 g / l; ■,50g / 1 [42].

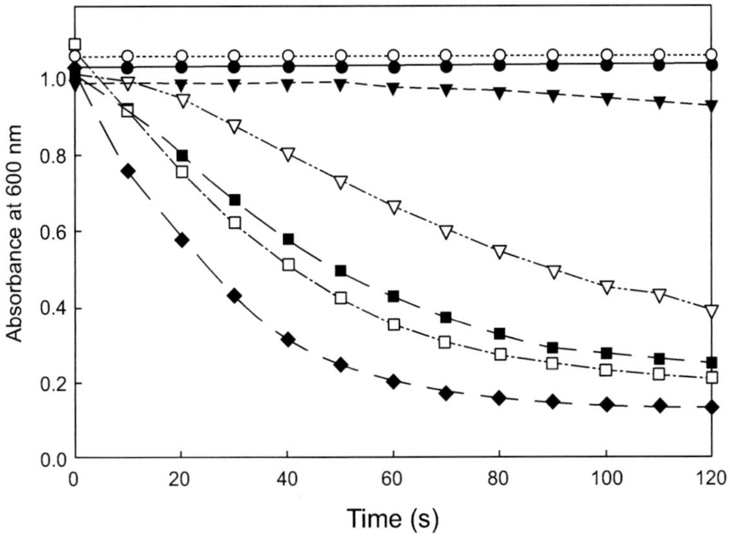

Fig. (13). Change in DyP activity under different initial molasses concentrations. Symbols: •,40 g/l ; o, 20 g/l ; ▼, 10 g/l ; ▽, 5 g/l ; ■,2 g/l ; □, 1g/l ; ◆, control (0 g/l) [42].

In contrast, freshly prepared molasses solution inhibited the activity of purified DyP toward RB5. Fig. (**13**) shows a change in DyP activity toward RB5 under different initial molasses concentrations. DyP activity decreased when the molasses concentration increased, and almost no DyP activity was observed at more than 10 g/l molasses concentration. The molasses was found to be effective in the production of DyP, as shown in Fig. (**12**), but it exerted inhibitory effects on the DyP activity, as shown in Fig. (**13**).

Fig. (**14**) shows the inhibitory effect of the two fractions of molasses prepared by ultrafiltration on DyP. When a 40 g/l molasses solution was fractionated into two fractions, both fractions completely inhibited the activity of DyP, as shown in Fig. (**14A**). When a 4 g/l molasses solution was fractionated into the two fractions, the fraction with a molecular mass less than 10 kDa exhibited stronger inhibitory activity than the other fraction, as shown in Fig. (**14B**). This means that inhibitory substances still existed in 4 g/l molasses solution and the lower molecular weight fraction especially showed stronger inhibitory effect on DyP.

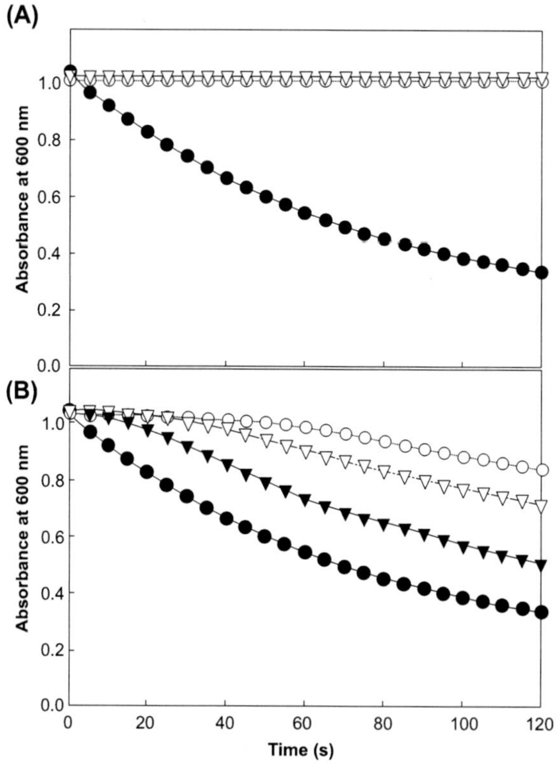

Fig. (14). Effect of fractionated molecular mass of molasses on the decolorizing activity of DyP toward RB5. (**A**) Molasses concentration of 40 g/l, (**B**) molasses concentration of 4 g/l. Symbols: ●,control; ○, non-fractionated molasses medium; ▼,fraction containing molecular mass of more than 10000; ▽, fraction containing molecular mass of less than 10000 [42].

Fig. (**15**) shows the change in the effect of 10 days of culture broth on DyP activity. The inhibition of the DyP activity decreased in the culture broth, suggesting that substances in the molasses involved in inhibition were degraded by Dec 1 during the 10-day cultivation. A similar result was reported by *Pleurotus eryngii* using methoxyhydroquinone or ferulic acid [123].

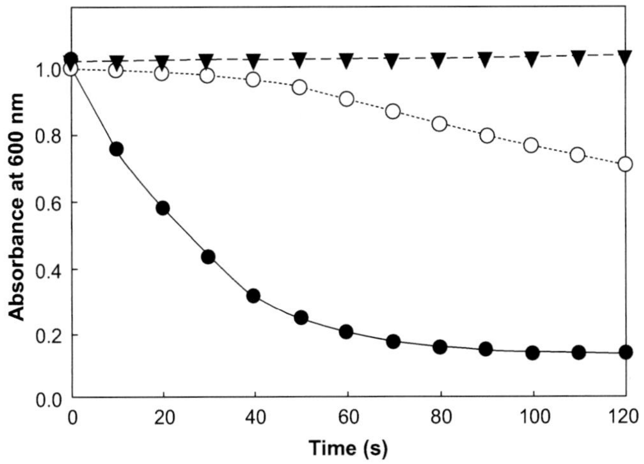

Fig. (15). Inhibitory effect on DyP activity of Dec 1 culture in 40 g/l molasses medium. Symbols: ●, control; ○, culture broth after 10 days; ▼, fresh molasses medium [45].

Fig. (**16**) shows gel chromatography patterns of molasses before and after cultivation of Dec 1, and the inhibitory effect of each fraction of fractionated molasses on DyP was assessed. The result that the absorbance of each fraction at 280 nm decreased after cultivation of Dec 1, as shown in Fig. (**16A**), means the change in the original black color of molasses to light yellow color by the Dec 1 consumption of molasses. Fig. (**16B**) shows the relative absorbance values of the decolorization activity of DyP. The 0-260 ml fractions did not exhibit a significant inhibitory effect because these fractions contained substances with a molecular mass of more than 80kDa in molasses. However, the 260-600 ml fractions containing a molecular mass of 1 to 80 kDa revealed strong inhibition against the activity of DyP, but the degree of inhibition was reduced during the growth of Dec 1. The 600-800 ml fractions with a molecular mass below 1 kDa showed greater inhibition by the fractions from the culture broth than that by the original molasses solution. However, absorbance at 280 nm of the culture broth after the cultivation was significantly decreased mainly because Dec 1 degraded large molecular mass substances.

Fig. (**17A**) shows the effect of dilution ratio of 10 day-cultivated cell-free culture broth on dye-decolorization activity. When the original culture broth decolorized RB5 at dilution ratio 1, the dye-decolorizing activity was about 50 mg/l/min. When the original culture broth was diluted at the dilution ratio of 25, the activity increased about 7 times higher than that of the original culture broth, and at dilution ratios of more than 25, the activity leveled off. This implies that the inhibitory effect of molasses can be eliminated at a dilution ratio of more than 25.

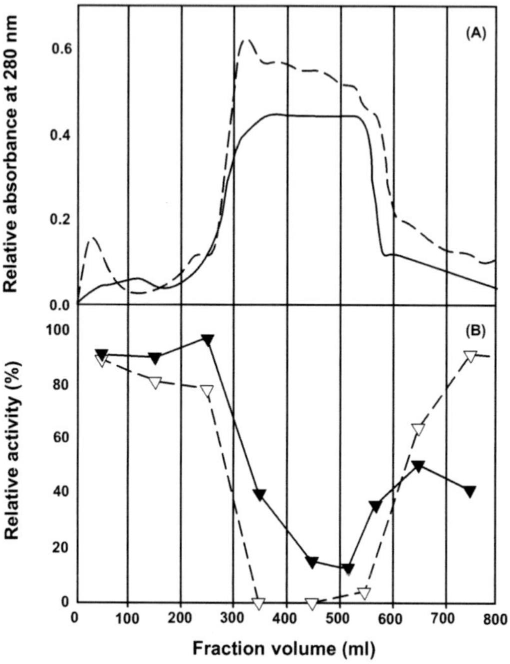

Fig. (16). Gel chromatography patterns of fractionated molasses before (------) and after (—) cultivation of Dec 1. The ordinates show relative values of absorbance at 280 nm (A) and relative decolorizing activity of DyP (%) (B). Symbols: ▼, after cultivation; ▽, before cultivation [45].

When the culture broth, presented in Fig. (**12**), was diluted 30 times, the dye-decolorizing activity was significantly enhanced, and the activity was the highest when 50 g/l molasses was used. At the significant inhibitory effect of molasses components shown previously, the values of decolorization activity were calculated (Fig. **12**) to be lower than the true values. At 10 and 20 g/l of molasses, a similar maximum activity was observed, as shown in both Fig. (**12**) and Fig. (**17B**). This indicates that the inhibitory substances in the molasses were completely degraded during the 10-day-cultivation of Dec 1.

The above results reveal that the molasses medium had two different effects on DyP production, namely stimulation of the production of DyP and inhibition of the activity of DyP. Therefore, the different effects should be taken into consideration for developing an optimal decolorizing process by determining the optimum molasses concentrations.

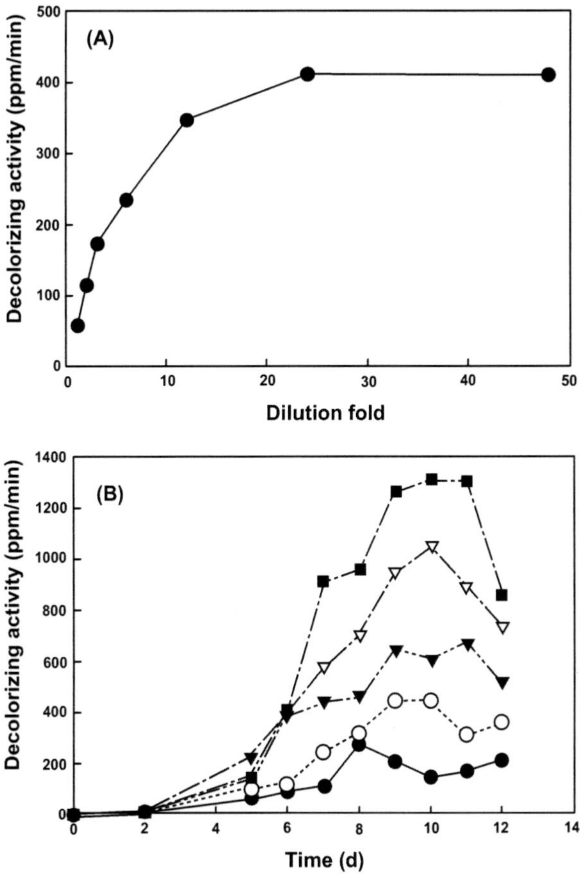

Fig. (17). Effect of dilution ratio of molasses concentrations on dye-decolorization activity of the cell-free culture broth derived from the 10-day culture of Dec 1 at the initial molasses concentration of 20 g/l (**A**). Effect of differently diluted molasses concentrations on calculated dye-decolorizing activities by 30 times diluted samples obtained in experiments in Fig. (**12**). (**B**) Symbols are the same as those shown in Fig. (**12**) [45].

In order to determine the positive effect of molasses on DyP production, process engineering was conducted following the two sequential reactor concepts for effective decolorization purposes. The flow rate of the molasses medium that was supplied to the first reactor had to be controlled at an optimum dilution rate for the Dec 1 growth and the DyP production. Then, the first reactor liquid was pumped to the second reactor and the dye solution was supplied to the second reactor at an optimum flow rate for decolorization. Through this hybrid process, the enhanced decolorization efficiency of both molasses and dye compensated for the disadvantage of using two reactors.

2. ARYL ALCOHOL OXIDASE (AAO)

2.1. Methods

2.1.1. Medium

A modified maltose BIII mineral medium that was applied for LiP production containing maltose as the sole carbon source [124] was used.

2.1.2. AAO Purification

Table **7** shows the conditions for AAO purification. The procedure has been described in detail in the previous paper [22].

Table 7. Purification steps for Aryl Alcohol Oxidase (AAO) [22].

	Volume	Protein	Total Activity	Specific Activity	Yield	Purification
	(ml)	(mg)	(U)	(U/mg)	(%)	(fold)
Supernatant	4045	203	24.4	0.12		1
YM 10	6.4	65	28.2	0.44	100	4
DEAE Sephacel	1.4	21	15.3	0.73	54	6
Butyl Toyopearl	1.1	4.0	4.7	1.2	17	10
Mono-Q	1.5	1.8	4.4	2.4	16	20

2.1.3. Detection of Maltose Concentration

The reducing sugar analysis method using the 3,5-dinitrosalicylic acid (DNS) was used [125].

2.1.4. Determination of Oxidization Activity of AAO in Cell-free Culture Broth

The cell-free culture broth was concentrated by ultrafiltration, and this solution and Veratryl Alcohol (VA) were mixed, and AAO activity was determined by measuring the oxidized product, veratraldehyde. The definition of one unit (U) of oxidization activity of AAO was the amount of AAO required to produce veratraldehyde at 1 μmol of VA per minute.

2.1.5. AAO Oxidizing Activity Toward Veratryl or Benzyl Alcohols

By using purified AAO, the following chemicals were oxidized: veratryl alcohol, anisyl alcohol, 4-hydroxy-3-methoxy benzyl alcohol, benzyl alcohol, and

4-hydroxy benzyl alcohol. The Km and Vm values of AAO were determined by Lineweaver-Burk plots.

2.1.6. DyP Decolorization Toward RB5 in the Mixture of DyP and AAO

RB5 was decolorized by DyP in the mixture of DyP and AAO to evaluate the contribution of H_2O_2 supplied from AAO to DyP, and the decolorization rate was measured by using the method described in the previous chapter [20].

2.1.7. Determination of H_2O_2 Concentration in the Culture Broth of Dec 1

The concentration of H_2O_2 during the cultivation of Dec 1 was determined by the oxidation of phenol red with Horseradish Peroxidase (HRP) [22].

2.1.8. Benzyl Alcohol and Aldehyde Derivatives Detection from Culture Broth of Dec 1

The amounts of benzyl alcohols of veratryl alcohol, anisyl alcohol, and 4-hydroxy-1-methoxy benzyl alcohol and their corresponding aldehyde derivatives were measured from the cell-free supernatant of Dec 1 culture using an HPLC column (Inertsil ODS-2) [22].

2.1.9. Prevention of Polymerization of Dye-degraded Products by AAO

The simplified model compound of RB5, AQ-2', was decolorized by the purified DyP mixed with AAO to determine the effect of AAO. In the first experiment, AQ-2', H_2O_2, and DyP were mixed. In the second experiment, veratryl alcohol and purified AAO were further added to the solution above. After both samples were lyophilized, the distribution of molecular weights of two samples was compared by gel permeation chromatography.

2.2. Results

2.2.1. Change in Activities of DyP and AAO

Fig. (**18A**) shows the change in RB5 decolorization rate by DyP and DyP and AAO activities in maltose BIII medium. Fig. (**18B**) shows the change in maltose concentration and dry cell mass of Dec 1. DyP activity was maximum in 8 days and disappeared in 12 days when no maltose was left in the medium. AAO activity increased even after maltose was exhausted, and the decolorization of RB5 lasted for 15 days mainly because polysaccharides were produced as by-products during maltose consumption by Dec 1. Polysaccharide production by Dec 1 has been confirmed previously in the growth phase [21]. A similar production pattern of AAO has been reported in *Pleurotus* species [126 - 128].

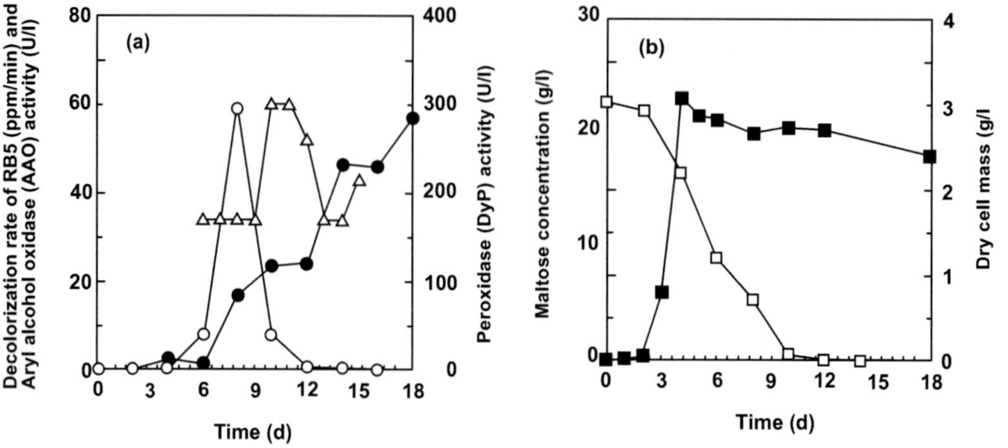

Fig. (18). Change in decolorization rate of RB5(Δ), AAO activity (•), and DyP activity (○) (**A**), and changes in maltose concentration (□) and dry cell mass (■) (**B**), produced during the growth of Dec 1 [22].

2.2.2. Purification of AAO

Table **7** shows the result for AAO purification. when Dec 1was cultivated in maltose III medium [22]. The molecular mass of AAO was 65.3 kDa. The specific activity of AAO was 2.4U/mg, which was found similar to 1.6 U/mg reported for *P. chrysosporium* [129].

2.2.3. Effects of pH and Temperature on AAO Activity

The optimal temperature for AAO activity was 45°C and treatment at 60°C for 30 min showed no residual activity. The residual activities of AAO at pH 6 after treatment at pH 3 and 7 were 12.5% and 96%, respectively. After treatment at 40°C for 30 min and at 60°C for 10 min, the residual activities at 30°C were 85% and 22%, respectively [22].

2.2.4. Substrate Specificity of AAO

Table **8** shows the kinetic parameters, Vmax and Km, of AAO for hydroxy and methoxy derivatives of benzyl alcohol derived from the Lineweaver-Burk plots. AAO did not oxidize benzyl alcohol and 4-hydroxy benzyl alcohol. Anisyl alcohol and 4-hydroxy-3-methoxy benzyl alcohol, methoxy derivatives of benzyl alcohol, were oxidized to their corresponding aldehydes. The highest oxidation rate of AAO obtained as Vmax was 54.1 U/mg for 4-hydroxy-3-methoxy benzyl alcohol.

Table 8. Kinetic parameters of aryl alcohol oxidase,AAO for several benzyl alcohol substrates[22].

Benzyl alcohol	λ^a(nm)	ε^b($M^{-1}cm^{-1}$)	K_m(mM)	V_{max}(U/mg)	V_{max}/K_m
Veratryl alcohol	310	$10.3x10^3$	1.7	4.2	2.4
Anisyl alcohol	290	$15.0x10^3$	0.5	19.7	39.4
4-Hydroxyl-3-methoxy benzyl alcohol	310	$8.8x10^2$	1.2	54.1	45.1
Benzyl alcohol	240	$9.6x10^3$	n.d.	n.d.	
4-Hydroxy benzyl alcohol	300	$8.9x10^3$	n.d.	n.d.	

a: Wavelength to shoiw each corresponding aldehyde derivative.
b: Molecular extinction coefficient at each λ.
n.d.: Not detectable.

2.2.5. Detection of Benzyl Alcohol, Aldehyde Derivatives, and H_2O_2 from the Culture Broth of Dec l

Fig. (**19**) shows the change in veratraldehyde concentration and H_2O_2 concentration detected during cultivation and decolorization of RB5 by Dec 1 in maltose III medium. Veratraldehyde was detected during decolorization of RB5 by the culture broth of Dec l. However, no veratryl alcohol was detected presumably because of the fast oxidation of extracellular veratryl alcohol to veratraldehyde by AAO. In separate experiments, the conversion of veratryl alcohol to aldehyde was found to be extremely fast.

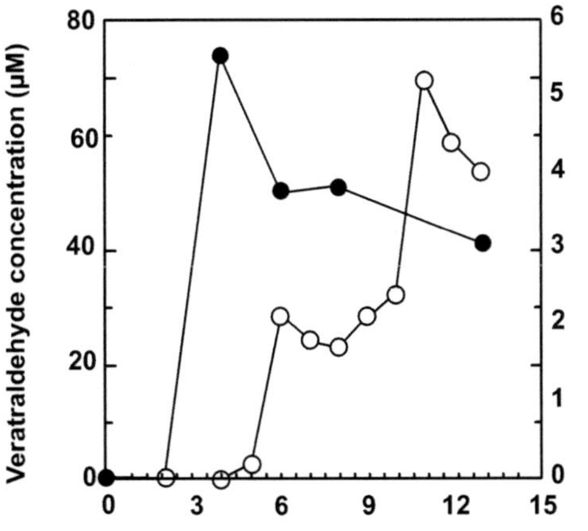

Fig. (19). Change in veratraldehyde concentration (•) and H_2O_2 concentration (o) [22].

Fig. (**19**) also shows the detection of H_2O_2 during cultivation of Dec l. Obviously, 1~2.5 µM H_2O_2 was constantly detected. As DyP demonstrated no ability to degrade lignin, AAO was supposed to play a role in supplying H_2O_2 to DyP. This indicates that the contribution of AAO to the decolorization of RB5 continued even after the activity of DyP was not observed after 12 days, as shown in Fig. (**18A**).

2.2.6. Dye Decolorization by the Mixture of DyP and AAO

The role of AAO in DyP reaction was investigated as follows: when AAO, Veratryl Alcohol (VA), and H_2O_2-free DyP were mixed, the dye-decolorization rate of RB5 by DyP was 0.57 μM /min. On the other hand, the decolorization rate of DyP with 0.2 mM H_2O_2 without AAO and VA showed to be 1.5μM/min. This suggests that H_2O_2 produced by AAO was utilized by DyP to decolorize RB5 during the growth of Dec 1.

2.2.7. Prevention of Polymerization by AAO

When DyP degraded 2,6-dimethoxyphenol, guaiacol, and several dyes *in vitro*, polymer production was observed. Actually, some polymers were precipitated when DyP decolorized AQ-2'. However, no polymer precipitation occurred when Dec 1 decolorized RB5. This suggests the possibility that AAO produced by Dec 1 suppressed the polymerization of RB5-derived products. Fig. (**20**) shows the molecular weight distribution of polymerized products. When AQ-2' was degraded by DyP, a polymerized product with a molecular weight of about 7500 was detected. When DyP and AAO were mixed, the molecular weight was distributed in the range of 665 to 7000. This indicates that AAO inhibited polymerization to a certain degree during the decolorization of AQ-2' by DyP.

From these results, it can be observed that during the decolorization of dyes by Dec 1, two kinds of roles of AAO became obvious, namely the supply of H_2O_2 to DyP and the depolymerization of dye-degraded products.

3. VERSATILE PEROXIDASE OF DEC 1, TCVP1

3.1. Methods and Results

3.1.1. Cultivation of Dec 1

The medium composition and culture procedure have been described in previous chapters [23].

3.1.2. Enzyme Purification

The activity of Manganese Peroxidase (MnP) was determined by forming Mn^{3+}–tartrate complexes from $MnSO_4$ in sodium tartrate buffer and H_2O_2. The activity of Versatile Peroxidase (VP) was determined by oxidizing guaiacol in sodium tartrate buffer and H_2O_2. Oxidation of guaiacol was supplemented with $MnSO_4$ to prove the effect of Mn^{2+} in the pH range of 2.5 and 5.5. The result represented Mn-dependent MnP activity.

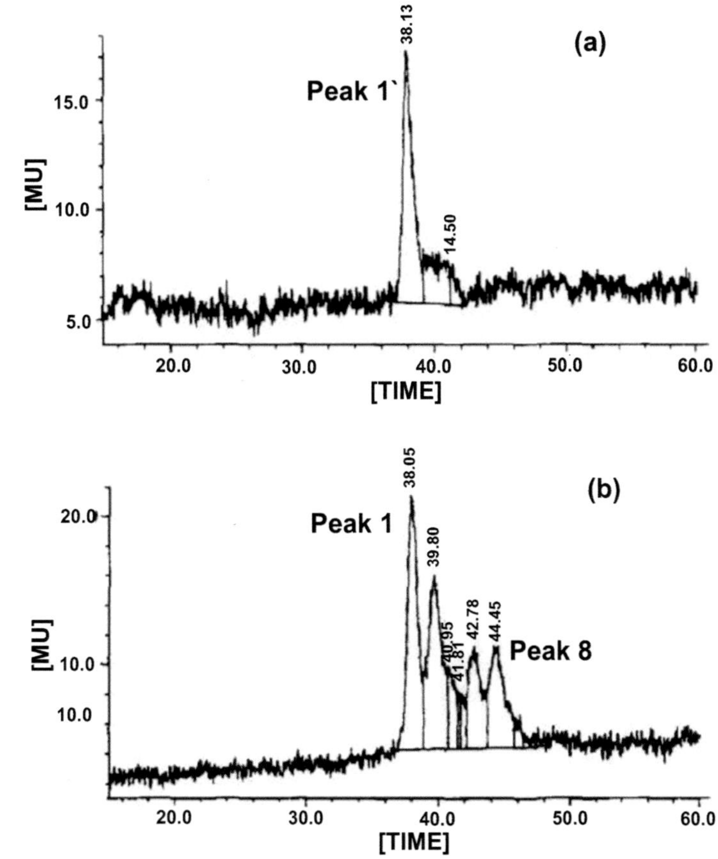

AQ-2

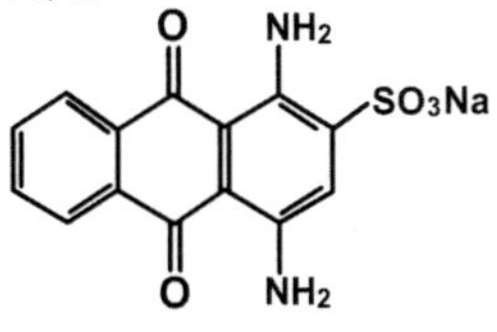

MW; 340 DA

Peak no.	Molecular weight (Da)	Time (min)
1`	7707	38.05
1	7568	38.13
2	3983	39.80
3	2487	40.95
4	2045	41.58
5	1811	41.87
6	1248	42.78
7	665	44.45
8		

Fig. (20). Distribution of molecular weight of polymerized products detected when DyP decolorized AQ-2' (**a**) and when the mixture of DyP and AAO decolorized AQ-2' (**b**). Peaks 1 to 8 correspond to peaks that appear on the charts in (a) and (b). The molecular weight and retention time of each peak are shown in the inset table [22].

The purification steps for Versatile Peroxidase (VP) produced by Dec 1, named TcVP1, are shown in Tables **9** and **10**. Total VP activity was about 30-fold lower than MnP activity, as shown in Table **10**. MnP activity was detected in each of the purification step, and the supernatant, ultrafiltration fractions, and QAE-Toyopearl fractions are shown in Table **9**. However, TcVP1 has shown MnP activity, and this activity was used to purify the protein by SDS-PAGE. A single band at 43 kDa was observed in the Mono Q chromatography fractions, as shown in Fig. (**21**). The value was estimated from standards for molecular mass estimation of myosin (200 kDa), β-galactosidase (116 kDa), bovine serum albumin (66 kDa), aldolase (42.4 kDa), and carbonic anhydrase (30 kDa). This value was almost the same as 40–48 kDa of other MnPs and VPs [83, 123, 127 - 131]. Although the MnP and VP activities were not separated clearly in the Mono Q chromatography fractions at elution, maximum MnP and VP activities were detected in 0.08 and 0.13 M NaCl elution fractions, respectively. Then, the fraction containing the maximum VP activity was further purified. The MnP and maximum VP activities appeared in fractions eluted with 80.5 and 100% Polybuffer 74, respectively. This indicates MnP and VP activities to have different Isoelectric points (p*I*).

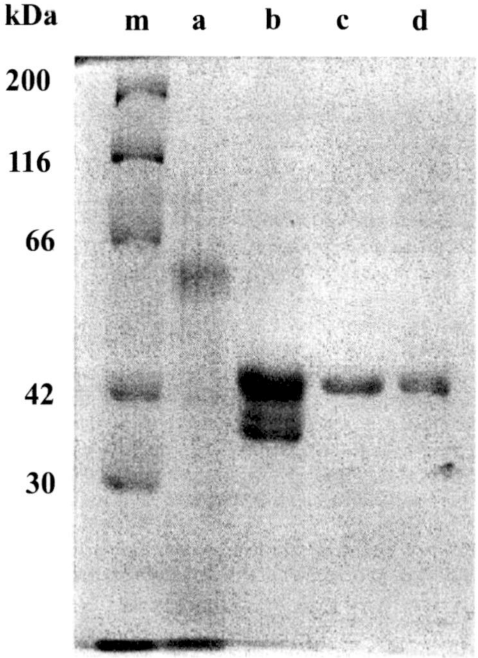

Fig. (21). Results of SDS-PAGE analysis of the fractions taken from each stage of TcVP1 purification as shown in Tables **9** and **10**. Lanes: m, molecular mass standards; a, ultrafiltration fraction; b, QAE-Toyopearl fraction; c, mono Q fraction; d, mono P fraction [23].

Table 9. Prepurification of TcVP1 (specific VP) from Dec 1 [23].

Purification Step	Total Protein	Total Activity	Specitic Activity	Recovery	Purification
	(mg)	(U)[a]	(U/mg)	(%)	(fold)
Culture supernatant	8.61	4.24	49	100	1.0
Ultrafiltration	6.53	3.23	50	76	1.0
QAE-Toyopearl fraction	2.27	225	99	53	2.0
Mono Q fraction	1.04	160	153	35	3.1
		[5.6]	[5.38]		

a: The listed activities are manganese oxidizing activity, except for the data shown in brackets, which represent guaiacol oxidizing activity (VP activity) in the absence of $MnSO_4$.

Table 10. Purification of TcVP1 from MnP by Mono P column chromatography [23].

Purification step	Total Protein	Total Activity	Specitic Activity	Recovery	Purification
	(mg)	(U)[a]	(U/mg)	(%)	(fold)
Mono Q fraction[b]	1.47	2.80	1.90	100	1.0
Mono P fraction for TcVPI[c]	0.36	2.67	7.41	95	3.9
For MnP	0.86	[99]	[119]	[97]	[1.2]

a: Guaiacol oxidizing activity (VP activity) in the absence of $MnSO_4$.
b: Mono Q fractions showing MnP/VP activity ratios < 50 were combined across five experiments.
c: TcVP1 and MnP fractions were separated by Mono P chromatography; data calculated from Mn oxidizing activity are shown in brackets.

3.1.3. N-terminal Amino Acid Sequencing of TcVP1

Table **11** shows the comparison of the results of N-terminal amino acid sequencing of the purified TcVP1 with those of other peroxidases. TcVP1 showed the highest homology with *T. versicolor* MPG 1, *T. versicolor* LiPG IV, *B. adusta* MnP1, and *Bjerkandera* sp. RBP with the matching level at 12 out of 14 amino acid residues and with an identity value of 86%. TcVP1 had moderate homology with *P. eryngii* MnPL1 and MnPL2 with matching level at 7 out of 14 amino acid residues and with an identity level of 50%. Relatively low homology was observed for *P. chrysosporium* MnP1 with an identity of 36% and *L. edodes* MnP and *P. chrysosporium* LiPH8 with an identity of 14%.

Dashes indicate gaps introduced in the probable alignment. X indicates an unknown residue. The amino acid residues that appeared in bold typeface were identical to those of TcVP1.

Table 11. N-terminal amino acid sequences obtained from reported peroxidases [23].

Sequences	Sources	References
VAXPDGVNTAXNAA	*Thanatephorus cucumeris* Dec 1 TeVP1	This study
VAXPDGVNTATNAA	*Bjerkandera adusta* DSM 11310 MnP1	[15]
VACPDGVNTATNAA	*Bjerkandera* sp. B33/3 RBP	[132]
VACPDGVNTAXNAA	*Trametes versicolor* MPG1	[133]
VACPDGVNTATNAA	*Trametes versicolor* LiPG IV	[133]
VTXS**D**TAVP	*Poriasubvermispore* MnP	[134]
ATCA**DG**RT**TA-NAA**	*Pleurotus eryngii* MnPL1	GenBank AF007221
ATCD**DG**RT**TA-NAA**	*Pleurotus eryngii* MnPL2	GenBank AF007222
AVC**PDG**TRV-SH**AA**	*Phanerochaete chrysosporium* MnP1	[28]
AVGS**DG**TVVP-DSV	*Lentinus edodes* MnP	[135]
ATCSNGK-VVPAAS	*Phanerochaete chrysosporium* LiPH8	[136]

3.1.4. Effects of pH and Temperature on the Activity of Versatile Peroxidase, TcVP1

The optimal pH determined in the oxidation of guaiacol for the Mn-independent and Mn-dependent activities of TcVP1 was 2.5 and 5.0, respectively, as shown in Fig. (**22**). These two activities were less than 5% of the maximum activity, indicating that at pH 5.0, TcVP1 reacted as a classical MnP-type, exhibiting indirect oxidation of guaiacol by TcVP1-generated Mn^{3+}. On the other hand, at pH 2.5, TcVP1 exerted direct oxidation of guaiacol by an Mn^{2+}-independent LiP-type reaction. From these data, the isolated TcVP1 was confirmed as a new type of VP.

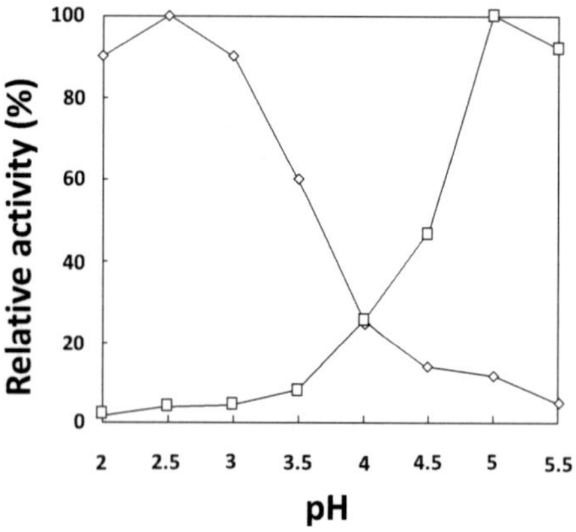

Fig. (22). Guaiacol oxidization activity by TcVP1 under different pH values. Symbols: □, 20 mM $MnSO_4$; ◇, no $MnSO_4$ [23].

Fig. (**23**) shows the effect of pH on the decolorization of RB5 by TcVP1 with or without Mn^{2+} supplementation. The decolorizing activity of TcVP1 toward RB5 was found to be higher with Mn^{2+} supplementation in the pH range of 2.0 and 5.0 than that without Mn^{2+} supplementation and maximum at pH 3.0, but it decreased sharply at more than pH3.0.

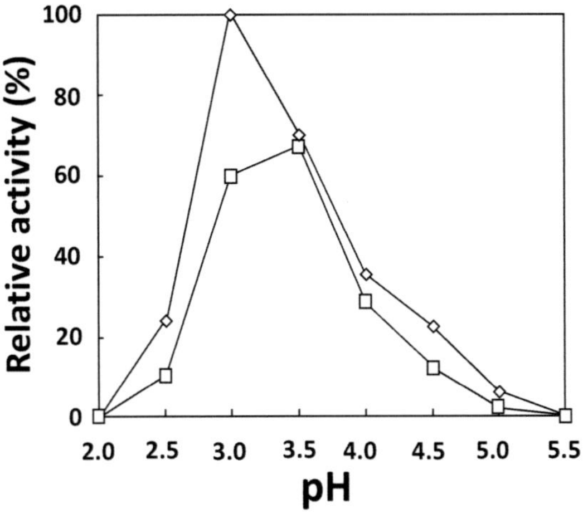

Fig. (23). RB5 decolorizing activity by TcVP1 under different pH values. Symbols:, ◇ 20 mM $MnSO_4$; □ no $MnSO_4$ [23].

3.1.5. Substrate Specificity

Table **12** shows a comparison of the substrate specificities for TcVP1 and recombinant DyP and rDy. rDyP exhibited higher decolorizing activity toward anthraquinone dyes, but relatively lower activity toward azo dyes. On the other hand, TcVP1 showed higher activity toward the azo dyes and significantly low activity toward anthraquinone dyes. TcVP1 also oxidized aromatic substrates, VA and guaiacol, as shown in Table **12**.

In the decolorization activity of two azo dyes, R. black 5 and R. red 33 without Mn^{2+} supplementation at pH 3, the specific activities of TcVP1 were significantly higher than those of rDyP, as shown in Table **12**. However, in the decolorization of anthraquinone dyes R. blue 5 and R. blue 19, the specific activities of TcVP1 were 50~60-fold lower than those of rDyP. From these results, it is obvious that the substrate specificity of TcVP1 toward azo was higher than that toward anthraquinone dyes. The substrate specificity of rDyP was opposite to that of TcVP1, and this led to complete decolorization of RB5 as sown in the next section.

Table 12. Comparison of substrate specificities for TcVP and rDyP [23].

Substrates	Chromophore	λ(nm)/ε ($M^{-1}cm^{-1}$)	Concentration (μM)	Specific Activity[a] of TcVP(U/mg)	Specific Activity of rDyP(U/mg)
VA		310/9,300	200	4	n.d.
Guaiacol		465/12,100	200	7.3	34
R. blue 5	AQ	600/8,000	125	6.9	340
R. blue19	AQ	600/16,000	50	4.3	270
R. blue 114	AQ	620/8,400	100	1.5	7.5
R. black 5	AZ	598/37,200	25	14.1	1.8
R. red 33	AZ	500/22,900	40	9.9	4.2
R. red 187	AZ	510/16,000	50	0.8	n.d.

VA: veratryl alcohol, AQ: anthraquinon AZ:azo, n.d.: not detectable.
a: The amount of each enzyme requirred to oxidize 1 μmol of substrate per minite at each reaction is defined as 1 U.

3.1.6. Complete Decolorization of RB5 by rDyP and TcVP1 In Vitro

Fig. (**24**) demonstrates spectra changes when rDyP decolorized RB5 *in vitro*. rDyP decolorized RB5 by 91% with an absorbance increase between 400 and 500 nm during the decolorization. As light red-brown products were generated, these products were designated as "colored intermediates". This phenomenon has been confirmed in a previous report [137]. When TcVP1 was added to the colored intermediates, complete decolorization occurred, as shown in Fig. (**25**). When Veratryl Alcohol (VA) was added at pH 3.0 to the colored intermediates reaction solution without Mn^{2+} supplementation, the decolorization ratio of RB5 was 86% of the initial absorbance after 12 h, as shown in Fig. (**25b**). Under the same conditions without VA addition, the decolorization ratio reached 76% after 24 h, as shown in Fig. (**25a**). As the decrease in the absorbance at 250~300 nm was significant after 12 h, this indicated the degradation of conjugated double bonds in the aromatic rings of RB5.

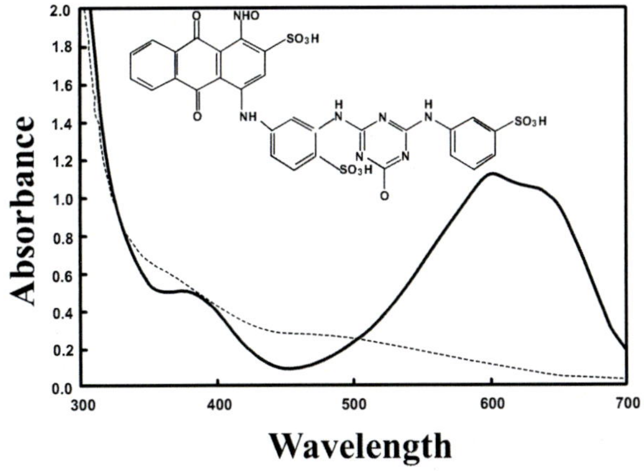

Fig. (24). Spectra change of RB5 when treated with or without rDyP. Solid line, intact RB5; broken line, RB5 treated with rDyP after 10 min. The molecular structure of RB5 is shown in Fig. (**1a**) [23].

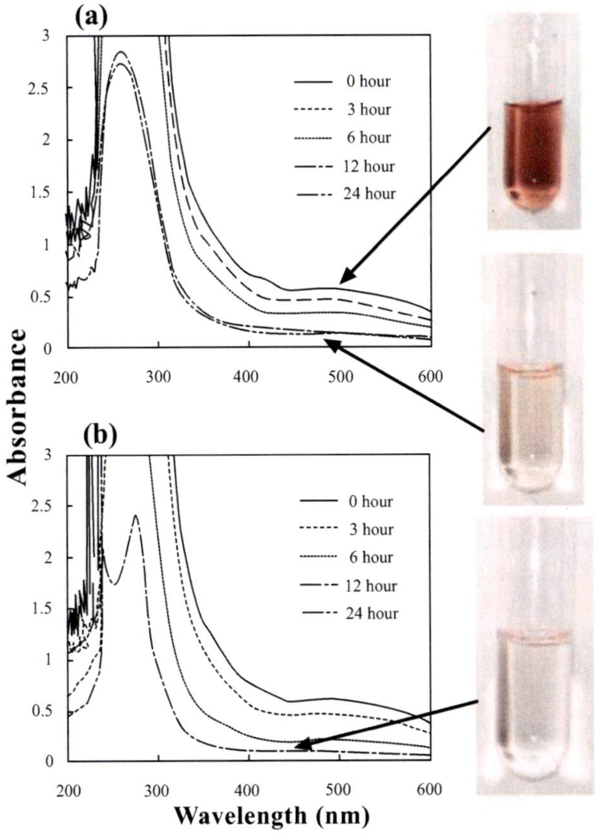

Fig. (25). Decolorization spectra of colored intermediates derived from TcVP1-treated RB5 at pH 3.0. **a**: Experiments without Mn^{2+} and veratryl alcohol, VA. **b**: Experiments without Mn^{2+} but with VA. The colors of pre- and post-reaction solutions are shown [23].

When Mn^{2+} was added in the experiment above, the decolorization ratio of RB5 was less than 20%. This indicates that Mn^{2+} inhibited the decolorization of RB5.

Complete decolorization of RB 5 with sequential addition of rDyP and TcVP1 is shown in Fig. (**26**).

Fig. (26). Decolorization process of RB 5 with rDyP and TcVP1 at 30°C. (the figure has been prepared by the author).

The reactants produced in the degradation of an anthraquinone dye, RB5, by rDyP were investigated, and a degradation pathway of RB5 by the rDyP has been proposed, as shown in Fig. (**27**) [138].

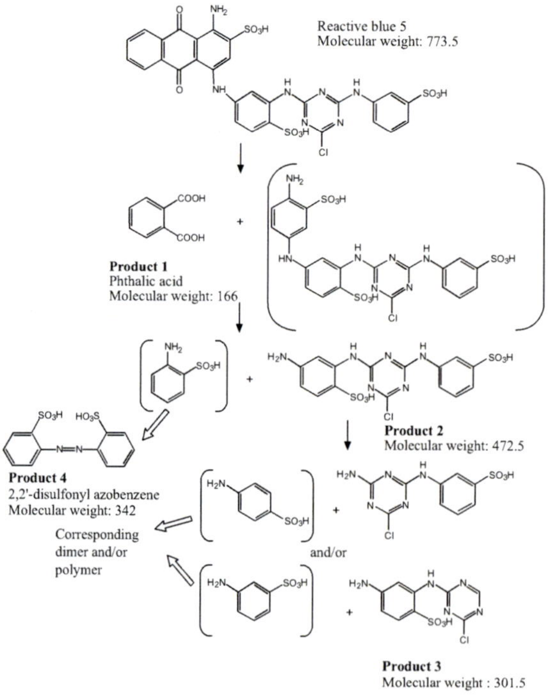

Fig. (27). Degradation pathway of RB5 treated with rDyP [138].

CONCLUSION

The three enzymes involved in the dye-decolorization of DyP, AAO, and TcVP1, were purified from the culture broth of Dec 1. A new peroxidase, DyP, comprising glycoprotein containing N-acetylglucosamine and mannose, degraded anthraquinone dyes effectively. A Soret band detected in the absorption spectrum of DyP proved DyP to be a hemoprotein. When $Na_2S_2O_4$ reduced the oxidized DyP, the presence of a protoheme was revealed as its prosthetic group.

Aryl Alcohol Oxidase (AAO) produced by Dec 1 was also purified. In the dye-decolorization culture of Dec l, H_2O_2 and veratraldehyde were detected in the presence of AAO. This implies that Veratryl Alcohol (VA) was produced by Dec l, and during oxidation of VA, generated H_2O_2 was supplied to DyP to support the dye-decolorization activity of Dec 1. AAO was also found to be involved in the depolymerization of the dye-degraded products.

A new versatile peroxidase, TcVP1, with a molecular mass of 43 kDa was purified. The result of amino acid sequencing of TcVP1 revealed the N terminus of TcVP1 to show high homology to several peroxidases produced by other fungi. TcVP1 preferentially decolorized azo dyes. Thus, the complete decolorization of RB5 proceeded sequentially; rDyP decolorized RB5 to light reddish-brown compounds, and then TcVP1 decolorized these colored intermediates to colorless. This is the first description of complete decolorization of an anthraquinone dye *in vitro*. The concerted action of DyP and TcVP1 has shown the mechanism of dye-decolorization by Dec 1, representing a novel strategy for the enzymatic treatment of dye wastewater.

The effect of molasses as a carbon source on Dec 1 activity and DyP production was analyzed. Decolorization activity by Dec 1 increased because molasses was used as a carbon source for the growth of Dec 1. However, certain molasses fractions separated by gel chromatography completely inhibited the purified DyP. These results indicate when applying molasses to the decolorization process, the process should be optimized by taking into consideration both the positive and negative effects of molasses.

CHAPTER 4

Enhanced Productivity of a New Peroxidase DyP by Genetic Manipulation and by Cultivation Methods

Abstract: A novel peroxidase, DyP gene, *dyp*, was cloned from a cDNA library of a newly isolated fungus, Dec 1. The open reading frame consisting of 1494 nucleotides indicated a primary translation product of 498 amino acids, and the Molecular mass (Mr) was estimated as 53,306. X-ray diffraction data using crystallized DyP revealed DyP to have a unique tertiary structure differing from that of most other well-known peroxidases. The analyzed amino acid sequence of DyP did not share homology with any other peroxidases except that of a peroxidase derived from *Polyporaceae* sp.

Mature cDNA encoding *dyp* was fused with the *Aspergillus oryzae* α-amylase promoter, *amyB*, and recombinant DyP was produced. The total activity of the purified recombinant DyP, rDyP, was produced. The total activity of rDyP, was about 400-fold higher than that of the native DyP derived from Dec 1.

The further productivity enhancement of rDyP was carried out using the following different cultivation methods and different media. Recombinant *A. oryzae* holding a DyP gene, *dyp*, was grown using the repeated batch method. When a synthetic liquid medium containing maltose as a carbon source was used in repeated batch culture, production of high-level rDyP activity continued for 26 repeated cycles of every 1-day batch.

When the production of rDyP by *A. oryzae* was carried out using complex media containing rice bran powder both in liquid repeated-batch and in fed-batch cultures, average rDyP productivities were similar in the two batch cultures.

The Solid-state Culture (SSC) was also attempted for the production of rDyP by recombinant *A. oryzae* using wheat bran as a solid medium and the productivity of rDyP was compared to that in the liquid cultures. The maximum productivity of rDyP in SSC reached 5.3g per kg wheat bran, and this productivity value was equivalent to the productivity using a 56 kg liquid culture. When the unit of the productivity per gram carbon of the medium was introduced, the productivity in SSC was 4.1-fold higher than that in the liquid cultures.

In order to overcome the disadvantages of SSC and liquid culture, Dec 1 was grown in an Air Membrane Surface (AMS) reactor. Although the growth of Dec 1 in AMS culture was almost the same as that in liquid culture at optimum temperatures, DyP productivity, DyP activity, and Aryl Alcohol Oxidase (AAO) activity in AMS culture

Makoto Shoda
All rights reserved-© 2024 Bentham Science Publishers

were 18-, 232-, and 108-fold higher than those in liquid culture, respectively. The advantage of AMS culture of Dec 1 1over SSC and liquid culture has been proven, particularly in the production of DyP.

A biofilm was formed in the AMS culture of Dec 1 and the relationship between biofilm formation and DyP and MnP was analyzed. Two new DyP isozymes, DyP2 and DyP3, were purified from the Dec 1 culture in the AMS reactor, and they were characterized and compared with the characteristics of rDyP.

Keywords: *Aspergillus oryzae*, Air-membrane surface reactor, Biofilm formation, Crystallization, Decolorizing peroxidase gene (*dyp*), DyP isozymes , Fed-batch culture, Recombinant DyP (rDyP), Repeated-batch culture, Rice bran powder, Solid-state culture, Wheat bran powder, X-ray diffraction, α-amylase promoter (*amyB*).

1. CLONING OF *DYP* GENE FROM DEC 1

The cloning and sequencing of gene, *dyp*, derived from Dec 1 cells are briefly described as follows [139]:

A peroxidase, DyP, derived from strain Dec 1 was partially digested by trypsin, and oligonucleotides encoding the two peptide fragments were synthesized for use as primers for Polymerase Chain Reaction (PCR). Then, the amplified DNA fragment by PCR was used as a probe for plaque hybridization using the cDNA library of Dec 1 derived from lambda phage λgt 10 [140]. The target gene was ligated to the plasmid pUC18 and sequenced.

Plaque hybridization produced eleven positive clones. The Molecular mass (Mr) of DyP was previously determined as 60,000. As this DyP contains 17%(w/w) saccharides, the Mr of the polypeptide was modified as 49,800 [20]. Therefore, an open reading frame corresponding to approximately 460 amino acid residues holding 1380 nucleotides was searched. PCR was carried out using the positive clones to estimate the size of each fragment and, one of them, clone 92 with1600 bp was digested into 2 fragments by *Eco*RI and was ligated into the *Bam*HI site of pUC18 and named pB92.

The pB92 contained an open reading frame consisting of 1494 nucleotides. The primary translation product of this open reading frame was composed of 498 amino acids and the molecular mass was estimated as 53,306. The value has been found to be reasonable with respect to the molecular mass of DyP. As all predicted amino acid sequences of the 5 oligopeptides obtained from DyP by trypsin digestion were found in this frame, the nucleotide sequence of this frame was judged to be the *dyp* gene encoding DyP in Dec 1.

2. CHARACTERISTICS OF DYP AS A PEROXIDASE

So far, all peroxidases have been categorized into animal peroxidase family and plant peroxidase family. The plant peroxidase family is further classified into three classes, classes I to III, according to the origin. As fungal peroxidases belong to class II peroxidases, DyP should belong to the class II peroxidases. However, the characteristics of DyP have been clarified to be significantly different from those of other known peroxidases [20, 139].

The first characteristic of DyP is as follows: the nucleotide sequence of the gene, *dyp*, and its primary translation product has no homology with any other peroxidases. One exception is a nucleotide sequence of peroxidase, *cpop*21 from *Polyporaceae* sp., which is registered under accession no. U77073 in GenBank, EMBL, and DDBJ. However, no characteristics of *cpop*21 have been published and thus, a detailed comparison between *cpop*21 and *dyp* is impossible.

The second characteristic of DyP is that all heme-containing peroxidases have two conserved His residues and one conserved Arg residue. One His residue is named proximal histidine and functions as the axial ligand for the heme. The other His residue is called distal histidine and Arg residue is known as essential arginine residue, being responsible for charge stabilization in the reaction between heme and H_2O_2 [20, 141, 142]. As all these three residues are conserved in all peroxidases, these residues have been considered essential for peroxidase activity, and there are no exceptions to this rule. Fig. (**28**) shows amino acid sequences of some well-known class II peroxidases together with DyP. It is clear that DyP has no heme-binding region, which is common to the plant peroxidase family [139]. Furthermore, His residue, which is responsible for heme binding, and is located at residue 51 of the amino acid sequence deduced from the cDNA of *dyp*, has been lost by N-terminal processing, as shown in Fig. (**28**). Therefore, DyP should have a heme-binding site different from those of other peroxidases, and thus the heme binding of DyP should be specific for Dec 1. Therefore, the recombinant rDyP obtained from *A. oryzae* should not express enzyme activity because it lacks heme. However, both original DyP and rDyP actually work as peroxidase. This suggests that there should be a novel heme-binding region other than the characterized region in the plant peroxidase family.

These are characteristic points of DyP as a new peroxidase.

```
                    □  ○                          ●
CCP I     43 GPVLVRLAWHTSG 55------165 REVVALMGAHALG 177
ECP I     97 AGLFIRMAWHGAG 109-----258 ETVALIAGGHTLG 270

ARP II    47 VRKILRIVFHDAI 59------175 DEVVDLLAAHSLA 187

MnP II    37 AHEVIRLTFHDAI 49------164 FEVVSLLASHTVA 176

LiP II    38 AHESIRLVFHDSI 50------167 LELVWMLSAHSVA 179

DyP II    42 QAPLPTLTQHTTE 54------165 VAPFTGTTIHGVF 175

TP  III   33 GASILRLFFHDCF 45------159 RDMVALSGAHTIG 171
HRP III   33 AASIIRLHFHDCF 45------161 SDLVALSGGHTFG 173
```

Fig. (28). Amino acid sequences of DyP and other peroxidases at regions of the conserved motif of the plant peroxidase family. The Roman numerals reveal the members of each class of the plant peroxidase family. The vertical lines indicate the identity between two peroxidases. The broken vertical lines indicate the matches between the residues classified according to the functional property of each amino acid. The number at both ends of each sequence means the number of residues from the initiation codon. CCP, cytochrome C peroxidase from *Saccharomyces cerevisiae*; ECP, peroxidase from *E. coli*; ARP, peroxidase from *Arthromyces ramosus*; MnP, manganese peroxidase from *P. chrysosporium*; LiP, lignin peroxidase from *P. chrysosporium*; TP, turnip peroxidase; HRP, horseradish peroxidase. Symbols, •, proximal histidine; ○, distal histidine; □, essential arginine [139].

3. EXPRESSION OF DYP IN *A. ORYZAE*

3.1. Characteristics of Recombinant DyP, rDyP

Cultivation of *A. oryzae* was described previously [139]. Table **13** shows the purification steps of rDyP produced by *A. oryzae* containing *dyp*. A molecular mass of rDyP was estimated as 58 kDa. The activity of purified rDyP was about 400-fold higher compared to that of the original DyP produced by Dec 1, although optimization of the production of rDyP was not conducted. As a Soret band at 407 nm reflected a hemoprotein, it became clear that rDyP used the same heme binding site as native DyP from Dec 1.

Table 13. Purification of rDyP from *A. oryzae* [128].

Purification Step	Protein	Total Activity	Sp Act	Yield	Purification
	(mg)	(U)	(U/mg)	(%)	(fold)
Culture supernatant	56.0	1.255	22.4	100	1
Centriprep	25.7	5.109	199	407	8.9
QAE-Toyopearl	14.0	3.241	231	258	10.3
Mono-Q	1.51	500	331	39.8	14.8

Sp Act: specific activity

Table **14** shows a comparison of characteristics between recombinant rDyP and native DyP.

Table 14. Comparison of characteristics between recombinant rDyP and native DyP [143].

Test or Characterisitic	Value for:	
	Native Dyp	rDyp
Substrate specificity (%) [a]		
Reactive blue 5	100	100
Reactive blue 19	92	100
Reactive black 5	0.5	0.9
Reactive red 33	2.0	1.8
Reactive yellow 2	2.6	0
Varatryl alcohol	0	0
Guaiacol	0.31	0.17
2,6-Dimethoxyphenol	0.61	0.54
Optimum pH [b]	3.2	3.0 - 3.2
Thermostability (%) at		
40 °C	97	100
50 °C	93	93
60 °C	79	12
Molecular mass(kDa) [c]	60	58

a: Relative activity toward RB5.
b: RB5 was used as the substrate.
c: Molecular mass was estimated from SDS - PAGE.

Almost no differences in substrate specificities for dyes between rDyP and native DyP were observed. The degradation activities of rDyP and native DyP toward veratryl alcohol, guaiacol, and 2,6-dimethoxyphenol were almost the same, but the thermostability and molecular mass between rDyP and native DyP showed a slight difference. The biochemical characteristics of DyP described in detail in previous sections [24] may also be adaptable to rDyP. As the N-terminal sequence of rDyP is identical to that of the amino acid sequence of native DyP [139], the processing sites of the N-terminal region of native DyP and rDyP would be the same. DyP is produced extracellularly, and thus a signal peptide of DyP secretion should exist in the N-terminal hydrophobic region. The hydrophobic N-terminal region of DyP was previously deduced to comprise around 20 amino acid residues from the initiation residue and that the potential cleavage site should be between residues 22 and 23 [139]. If this is also adaptable for rDyP, the processing may occur in a similar way in both fungi of Dec 1 and *A. oryzae*. Although the two fungi belong to different genera, the biochemical process of DyP depends solely on DyP itself, and not on the genus of fungi.

The small difference in Molecular mass (Mr) of rDyP and native DyP, corresponding to 58 kDa and 60 kDa, respectively, can be mainly attributed to the difference in the glycosylation ratio between the two DyPs.

3.2. Crystallization and X-ray Analysis of DyP

The purified DyP was crystallized and then X-ray diffraction analysis was conducted. The result of the crystallized DyP revealed that the crystal belonged to the hexagonal space group P6522 [144]. From the X-ray diffraction data, DyP has been found to contain four-stranded antiparallel β-sheets, and this structure was significantly different from that of conventional LiP and MnP. Furthermore, the arrangement on the distal side of the heme pocket was clarified to be significantly different from the structures of LiP and MnP [24, 63, 144, 145].

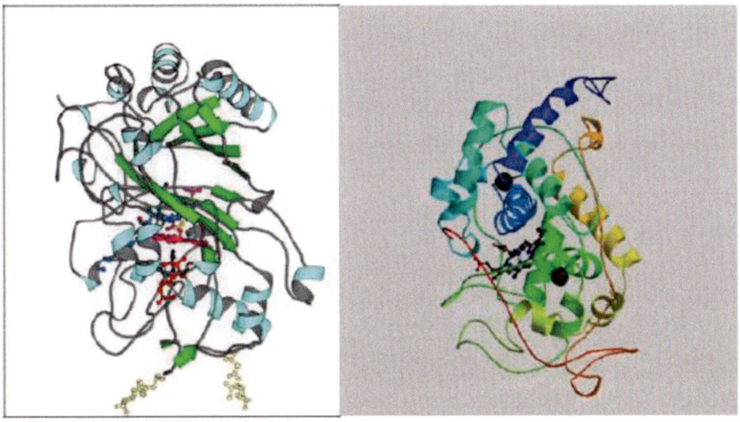

Fig. (29). Structure of DyP (left) from Dec 1 and lignin peroxidase from *Phanerochaete chrysosporium* (right) [24, 144].

Fig. (**29**) shows the different structures of DyP and lignin peroxidase from *P. chrysosporium*.

As it has been proven that DyP is a novel peroxidase that forms an independent family separated from the classical peroxidase groups, DyP has been designated as "DyP-type peroxidase family", and some peroxidases have been re-categorized to DyP-type peroxidase family, as described in Chapter 1.

4. ENHANCED PRODUCTION OF DYP OR RDYP

4.1. Repeated-batch and Fed-batch Cultures by Recombinant *A. oryzae*

4.1.1. Methods

4.1.1.1. Strain

The recombinant *Aspergillus oryzae* RD005 carrying the *dyp* gene was used. The cultivation method of *A. oryzae* RD005 has been described previously [139].

4.1.1.2. Batch Culture

After recombinant *A. oryzae was* grown on maltose agar plates, a spore suspension obtained from the cell suspension was put into flasks containing 1% wheat bran powder or rice bran powder as a main substrate, and batch culture was started.

4.1.1.3. Repeated-batch Culture

After recombinant *A. oryzae* was cultivated in batch culture by the method described above, the repeated-batch culture was started as follows: when rDyP activity reached the maximum, the cells of *A. oryzae* were statically sedimented, and the culture supernatant was discarded. Then, the same amount of fresh wheat bran powder or rice bran powder media was added and the next repeated-batch cultivation was started. This procedure was repeated.

4.1.1.4. Fed-batch Culture

After the batch culture described above, the fed-batch culture was started as follows: wheat bran powder or rice bran powder was added every other day for 11 days as the first fed-batch culture. Then, the cells in the culture were sedimented to remove the cell-free culture, fresh wheat bran powder or rice bran powder media was supplied, and the second fed-batch culture was started.

4.1.1.5. Media

The six different media were prepared to compare DyP productivity. Table **15** shows the compositions of these media and other inorganic compounds. When the wheat bran powder or the rice bran powder were used as main substrates, they were named as medium-1 and medium-3, respectively.

According to a previous study by *A. oryzae*, enzyme production increased significantly in phosphate-supplemented soybean meal medium [147]. Therefore, the effect of phosphate supplementation was investigated in repeated-batch cultures using wheat bran powder and rice bran powder, named medium-2 and medium-4, respectively. In fed-batch culture experiments, the sterilized rice bran powder and sterilized wheat bran powder were used, named medium-5 and medium-6, respectively.

Table **16** shows the chemical compositions of the wheat bran powder and rice bran powder.

Table 15. Composition of four media used in repeated-batch cultures (g/l) [146].

Media	K_2HPO_4	KCl	$MgSO_47H_2O$	WBP	RBP
Medium-1	1	0.5	0.5	15	
Medium-2	1.5	0.5	0.5	15	
Medium-3	1	0.5	0.5		15
Medium-4	1.5	0.5	0.5		15

WBP:Wheat bran powder
RBP: Rice bran powder

Table 16. Composition of rice bran and wheat bran powders [131].

Substrate	Component (%)						
	Water	Ash	Protein	Lipid	Carbohydrate	Cellulose	Total Phosphate
Wheat bran powder	14	4	16	3	63	-	1.85
Rice bran powder	13.5	8.9	13.2	18.3	38.3	7.8	4.73

—, not measured.

4.1.1.6. Determination of rDyP Activity

rDyP activity was determined using the mixture of the culture supernatant, dye, RB5, and H_2O_2 by employing the method described previously [67, 143]. The definition of one unit of DyP activity was the amount of DyP required for the decolorization of 1 μmol of RB5 in 1 min.

The average productivity of rDyP was calculated using the following equation [146]:

Average productivity (U/ml/d) = ∑increase in rDyP activity in each batch or fed-batch (U/ml)/incubation time (d)

The maximum productivity of rDyP was the maximum value of rDyP detected during repeated-batch cultures or fed-batch cultures. The incubation time included the time of the initial batch culture.

4.1.2. Results

4.1.2.1. Repeated-batch Culture Using Medium-1

Fig. (**30**) shows the changes in pH and rDyP activity during repeated-batch culture of medium-1 containing wheat bran powder. rDyP activity reached the maximum of 3~5 U/ml during the four repeated batches, reflecting that the cells of *A. oryzae* reached the maximum within 2 days. As rDyP was inactivated at pH values above 7 [67], the pH values of the cultures were controlled around 7. On day 13, the medium-2 containing an increased concentration of phosphate was supplied, as indicated by white arrows. Then, rDyP production was enhanced and the recycle time decreased to only 1 day compared to 2 days in medium-1. The maximum productivity of rDyP, 6 U/ml/d, was assessed in the 7th repeated batch culture, and the average productivity during 17 days was 2.8 U/ml/d.

4.1.2.2. Repeated-batch Culture Using Medium-3

Fig. (**31**) shows the result using medium-3. Maximum rDyP activity was 6.4 U/ml in the third repeated culture. Medium-4 containing additional phosphate was supplied on day 14. However, phosphate supplementation did not increase DyP activity but decreased rDyP activity. As the phosphate content of rice bran powder was 4.73%, as shown in Table **16**, which was more than 2 times higher than that in the wheat bran powder, the additional phosphate rather adversely affected rDyP production in rice bran powder. The average rDyP productivity was 2.3 U/ml/d.

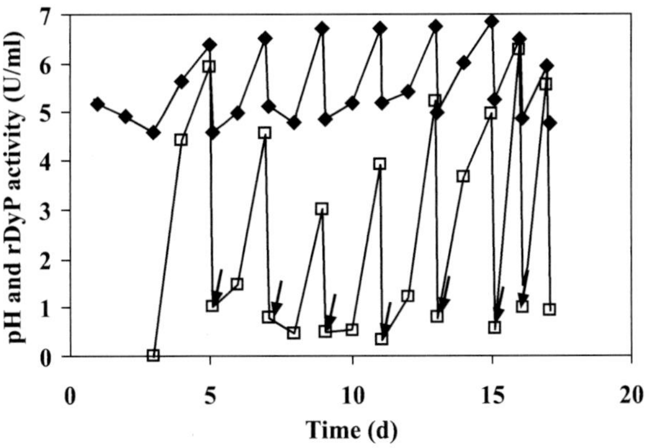

Fig. (30). Changes in pH (◆) and rDyP activity (□) in medium-1 supply, indicated as black arrows. White arrows indicate medium-2 supply [146].

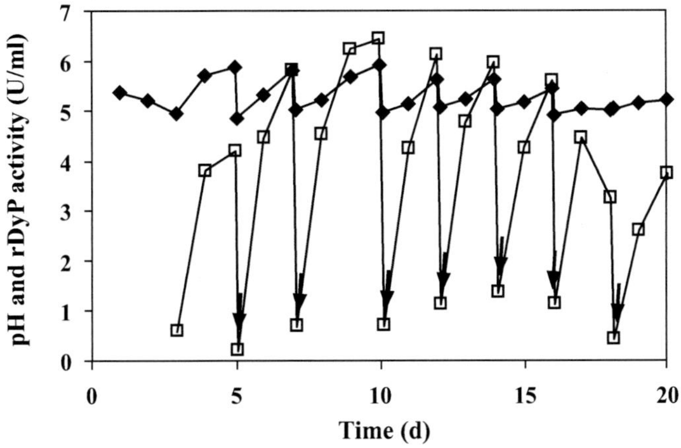

Fig. (31). Changes in pH (♦) and rDyP activity (□) in medium-3 supply, indicated as black arrows. White arrows indicate medium-4 supply [146].

4.1.2.3. Fed-batch Culture Using Medium-5

Fig. (**32**) shows the result using rice bran powder as medium-5. When rDyP activity reached 4 U/ml in 5 days, fed-batch culture was started by supplying medium-5. rDyP activity reached 20 U/ml in 11 days. At this time, rice bran brown particles and mycelia of Dec 1 formed clumps and deficient dissolved oxygen occurred. Then, the supernatant obtained after the cells were sedimented was replaced with medium-4 and the second fed-batch culture mode was started by adding medium-4 every other day. Finally, rDyP activity was obtained as 18.8 U/ml. The average rDyP productivity in the first fed-batch culture and the second culture was 1.7 U/ml/d and 2.1 U/ml/d, respectively. The average rDyP productivity was 1.9 U/ml/d.

4.1.2.4. Fed-batch Culture Using Medium-6

Fig. (**33**) shows the result using medium-6. When rDyP activity was 11.7 U/ml on day 11, medium-2 was supplied and the second fed-batch culture was started after day 11 by adding medium-2. However, rDyP activity started to decline, and became 3.5 U/ml after 17 days. The average rDyP productivity in the first fed-batch culture and the second fed-batch culture was 2.7 U/ml/d and 1.2 U/ml/d, respectively. This lower productivity was presumably due to a lower content of phosphate in medium-6.

4.1.3. Discussion

The results of the four cultures are summarized in Table **17**. Generally, it is difficult to determine the fungal cell masses especially when solid-containing complex media, like wheat bran powder and rice bran powder, are used because the solid materials interfere with accurate measurement of cell mass. Therefore, the same amounts of the two solid-containing media were added in the cultures, and differences in activity and production of DyP between the two materials were compared.

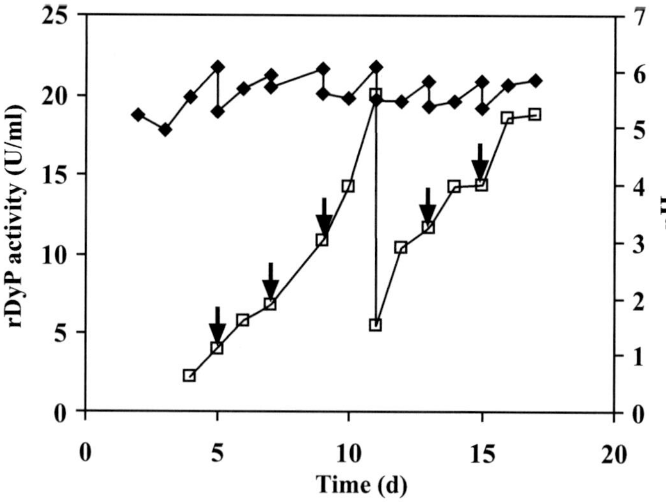

Fig. (32). Changes in pH (◆) and rDyP activity (□) during fed-batch culture using medium-5. The arrows indicate intermittent supply of medium-5 before day 11 and medium-4 after day 11 [146].

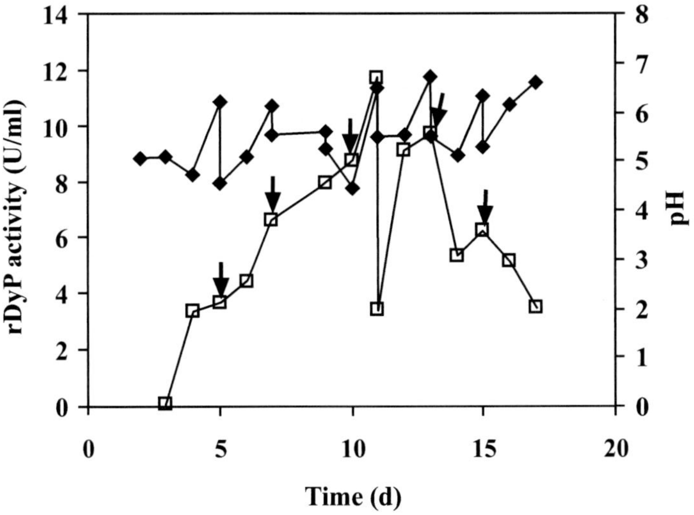

Fig. (33). Changes in pH (◆) and rDyP activity (□) using medium-6. The arrows indicate an intermittent supply of medium-6 before day 11 and medium-2 after day 11 [146].

Table 17. Results for each culture [131].

	Repeated-batch Culture		Fed-batch Culture			
	Wheat Bran Powder	Rice Bran Powder	Rice Bran Powder		Wheat Bran Powder	
Total amount added (g) (A)	12	12	6[a]	4.5[c]	6[a]	4.5[c]
Total rDyP activity (U) (B)	3920	4190	2000	1875	1170	356
Yield = B/A	326	349	333	417	195	79
Average productivity of rDyP (U/ml/d)	2.8	2.3	1.7 (2.4)[b]	2.1 (1.9)[d]	1.2 (2.7)[b]	-[e]

a: First fed-batch culture.
b: Average excluding the first batch culture.
c: Second fed-batch culture.
d: Average over the whole incubation period.
e: Not calculated.

The total rDyP activities produced and the average rDyP productivities in repeated-batch cultures using the two media were not different, as shown in Table **17**, suggesting similar growth of Dec 1 cells in the two media.

Supplementation of phosphate accelerated enzyme production and conidiophore formation by *A. oryzae* cultivation using soybean meal reported previously [137]. In this Dec 1 case, the effect of phosphate was also confirmed, showing that rDyP productivity increased in the repeated-batch culture of wheat bran powder medium.

As enzymes are generally produced as secondary metabolites, the process of metabolism is associated with fungal sporulation or conidiation [148, 149]. Fig. (**34**) provides a microscopic observation of recombinant *A. oryzae* cultivated in wheat bran powder medium containing different phosphate concentrations. It became clear that by increasing the phosphate, the number of conidia increased, resulting in the enhanced production of rDyP. However, rice bran powder did not show any positive effect of phosphate presumably because sufficient phosphate content of 4.73% in rice bran powder caused detrimental protease production, inhibiting the activity of rDyP or causing the accelerated aging of microorganisms [35, 150].

Fig. (**30**) shows a rapid pH increase in the repeated-batch culture of wheat bran powder, leading to a 1-day cycle of repeated-batch culture to maintain pH in the optimum range [67]. On the other hand, in rice bran powder medium, fairly stable pHs during the two batch cultures needed a longer cycle of culture, as shown in Figs. (**31** and **32**). This means that specific attention to pH was not needed in this case.

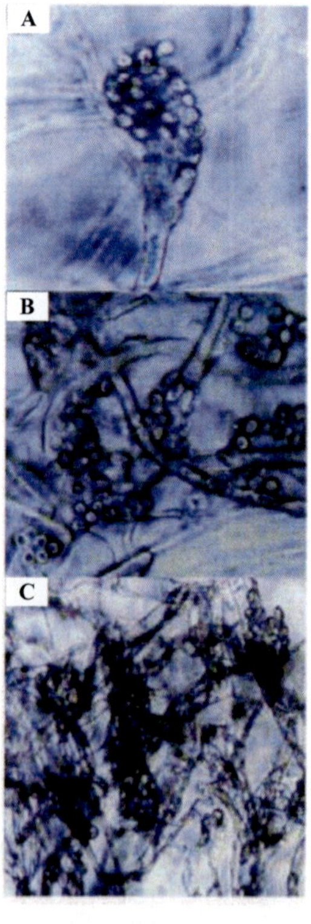

Fig. (34). Changes in conidia formation of recombinant *A. oryzae*. grown in wheat bran powder medium at different K_2HPO_4 concentrations of 0.02% (**A**), 0.05% (**B**), and 0.1% (**C**) [146].

The average rDyP productivity in rice bran powder medium was similar in the repeated- and fed-batch cultures, suggesting that *A. oryzae* led to similar production of rDyP. Contrarily, rDyP production by *A. oryzae* in wheat bran powder medium was significantly different in the two batch cultures presumably because of insufficient phosphate concentrations in the wheat bran powder. Therefore, phosphate was found to be a key factor when wheat bran powder was used as a substrate.

In section 4-3, the use of Solid-state Culture (SSC) will be discussed in detail. A previous report described that SSC using natural complex substrates, like wheat bran powder and rice bran powder, provided better enzyme productivity than liquid culture [151]. However, careful comparison is needed regarding the productivity between SSC and liquid culture. Our previous report presented quantitative data on the two cultivation methods based on medium weight or culture volume, and showed that the maximum rDyP productivity in liquid culture using synthetic medium and SSC using wheat bran powder was 5.25 U/d/g of medium and 219 U/d/g of medium [67], respectively. In the present study, the maximum rDyP productivity in repeated-batch culture using wheat bran powder was 6 U/ml/d, and this value was converted to 342 U/d/g of the medium. This value was calculated as 65- and 1.6-fold higher than the previous values in the liquid and SSC experiments, respectively. This indicates that the repeated-batch culture conducted here can be a better method for rDyP production than SSC because system control and operation of liquid culture are relatively easier in this method than in SSC.

Several previous researches have shown enhanced MnP productivity results in the repeated-batch process. Table **18** shows the results of previously reported MnP productivity together with that of rDyP. rDyP showed significantly higher enzyme activity than the previously reported data. Actually, the results in Table **18** cannot be compared directly because of different factors involved, such as reactor configurations, operation modes, fungal species, substrates, and unit enzyme definition. The advantages of the present method include shorter incubation times and the use of inexpensive complex substrates.

Immobilization of fungal cells has been reported to be efficient in retaining the cells in a reactor homogeneously and reusing fungal cells [152 - 154]. Actually, the immobilization of fungal cells in large-scale reactors has several disadvantages, such as laborious cell immobilization, limited oxygen transfer, disruption of the immobilized matrix by agitation and aeration, and inactive spore formation [155]. However, as elucidated in Chapter 6, Dec 1 immobilization on polyurethane foam was successful especially in repeated-batch mode operation because of the advantages of an increase in DyP production, which facilitated the separation of the cells from the culture supernatant and simplified reuse of the cells. Furthermore, immobilized cells reduced the viscosity of the culture and improved the mass transfer of oxygen and nutrients [156]. Thus, repeated-batch culture is more applicable and effective than other processes when immobilized cells of recombinant *A. oryzae* are used [157 - 159].

Table 18. Summary of published results for manganese peroxidase production by repeated-batch fungal cultures [146].

Configurations	Cycle Time (d)	Substrate (g/l)	Fungus Species	Average Peroxidase Productivity (U/l/d)	References
Repeated-batch	1-2	WBP, 10	*A. oryzae*	2800	This study
	2-3	RBP, 10	*A. oryzae*	2300	
Fed-batch	7-11	RBP, 10 (40[a])	*A. oryzae*	1700	
	7-11	WBP, 10 (40[a])	*A. oryzae*	1200	
Repeated-batch (immobilized)	2-4	Glucose, 1	*P. chrysosporium*	239	[137]
Repeated-batch	1-3	Glucose, 5 Na acetate,2.7	*N. frowardi*	246	[139]
Repeated-batch (immobilized)	7.5-10	Glucose, 10	*P. chrysosporium*	46	[35]
Repeated-batch (immobilized)	5	Glucose, 13	*C. dusenii*	439	[138]

WBP, Wheat bran powder, RBP, rice bran powder.
a Cumulative values.
b One unit of MnP activity was defined as the amount of enzyme required to produce 1 μmol of Mn3 - malonate in 1 min.
c One unit of MnP activity was defined as the amount of enzyme required to oxidize 1μmol of 2,6-dimethoxyphenol in 1 min.

4.2. Repeated-batch Liquid Cultivation by Recombinant *A. oryzae* in Synthetic Medium

4.2.1. Methods

Recombinant *A. oryzae* was cultivated in the liquid medium under different conditions, as shown in Table **19**. As a dye, the anthraquinone dye, Remazol Brilliant Blue R (RBBR), was selected. By mixing *A. oryzae* and RBBR, dye decolorization was initiated by adding H_2O_2, and rDyP activity was measured in the repeated-batch mode [160].

When the batch culture in flasks continued until the rDyP activity reached more than 2 U/ml, the repeated-batch culture was started as follows: on each day, the cells of *A. oryzae* were statically sedimented to the bottom of the flasks and the supernatant was removed. Then, the same amount of fresh repeated-batch culture medium was introduced and cultivation was continued. In the first six cycles of repeated-batch culture, the medium contained 0.1% (w/v) $NaNO_3$, and in the subsequent cycles, 0.05% (w/v) $NaNO_3$ was added because nitrogen limitation improved the production of DyP, as also observed in a previous experiment [1]. Maltose was selected as a carbon source because in preliminary experiments using several sugars, maltose showed high production of rDyP. Maltose induced Taka-amylase A of *A. oryzae* that was transformed with the gene *dyp* from Dec 1 using the Taka-amylase promoter of gene, *amyB* [143].

Table 19. Compositions of preculture, batch culture, and repeated-batch culture media (%) (w/v) for recombinant *A. oryzae* cultivation [160].

Culture	$NaNO_3$	K_2HPO_4	$MgSO_4 \cdot 7H_2O$	KCl	$FeSO_4 \cdot 7H_2O$	Maltose	Yeast Extract	$CaCO_3$
Preculture	0.2	0.1	0.05	0.05	0.002	3	1	
Batch culture	0.2	0.1	0.05	0.05	0.002	3		0.1
Repeated-batch culture	0.05-0.1	0.1	0.05	0.05	0.002	1		0.1

The average productivity of rDyP (U/ml/d) was defined as the sum of the increase in rDyP activity in each batch (U/ml) divided by incubation time (day).

The RBBR dye's decolorization ratio (%) was obtained as follows:

$$[(Ci — Cf)/Ci] \times 100$$

The apparent decolorization rate was calculated by:

$$(Ci — Cf)/t$$

Where Ci and Cf are the initial and final concentrations of the dye, respectively, and t is the time.

rDyP activity analysis has been described in detail in previous chapters. Thus, a brief description is given as follows: the supernatant of the samples was used for rDyP activity analysis by mixing H_2O_2 and anthraquinone dye, RB5, and decolorization was monitored spectrophotometrically. The definition of one unit of the rDyP activity is the amount of the rDyP required to decolorize 1 μmol of RB5 in the reaction mixture in 1 min.

4.2.2. Results

Fig. (**35**) shows changes in the rDyP activity and pH both in the initial batch culture for 6 days and in consecutive repeated-batch cultures. After 6 days, the rDyP activity reached 2.6 U/ml, and daily repeated-batch cultures were started using fresh $NaNO_3$ and maltose. During cultures, *A. oryzae* formed pellets, and the pH was maintained lower than 7 to secure rDyP activity [67]. Then, 0.05% (w/v) $NaNO_3$ was added to reduce nitrogen concentration to maintain more than 3 U/ml of rDyP activity until 31 days. Then, in order to observe a low-temperature effect of *A. oryzae* pellets, *A. oryzae* pellets that were kept in fresh repeated-batch culture medium were maintained at 4°C for 20 days. Thereafter, the sample was returned to the normal active culture conditions and rDyP production and its activity were measured. Interestingly, rDyP production resumed within a day and the activity reached 4.6 U/ml, as shown in Fig. (**35**). The original DyP activity of Dec 1 was approximately 0.001 U/ml, and the final activity of rDyP was 4.6 U/ml, showing an approximately 4000-fold increase. The average productivity of rDyP was 2.82 U/ml/d. This repeated batch mode and stable pH control were closely related to high rDyP production [161, 162].

So far, laccase was the only enzyme reported to decolorize RBBR [137, 163 - 165]. Here, rDyP obtained showed to be an effective enzyme for the dye decolorization.

The repeating cycle time is crucial in the repeated-batch culture for efficient reactions [166]. The one-day cycle time shown here was the shortest among other repeated-batch cultures for enzyme production [153, 154].

The cumulative dye and H_2O_2 concentrations and residual dye concentration with the intermittent supply of the dye and H_2O_2 are shown in Fig. (**36**). The apparent decolorization rate was calculated as 17.7 mg/l/min from the data obtained in Fig. (**36**).

The cost of synthetic maltose medium used in this study was relatively high compared to that of complex media, like wheat bran and rice ban powders, used previously. However, the cycle number of repeated culture was significantly longer in the synthetic maltose medium, leading to higher production of rDyP. Therefore, the process must be optimized by considering the advantages of the two media.

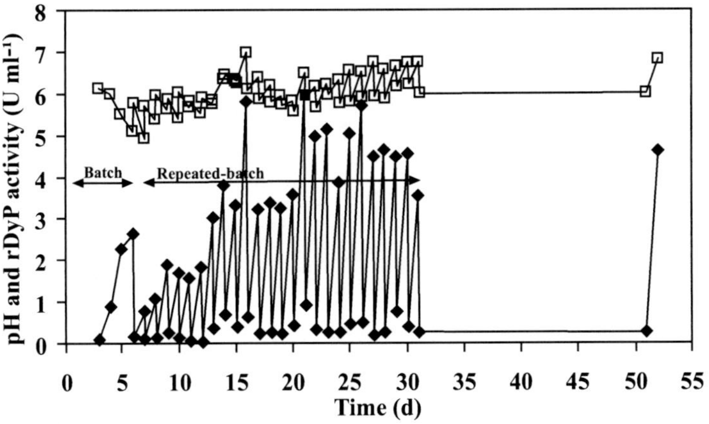

Fig. (35). Changes in rDyP activity (◆) and culture pH (□) during an initial batch and consecutive repeated-batch culture of *A. oryzae* [160].

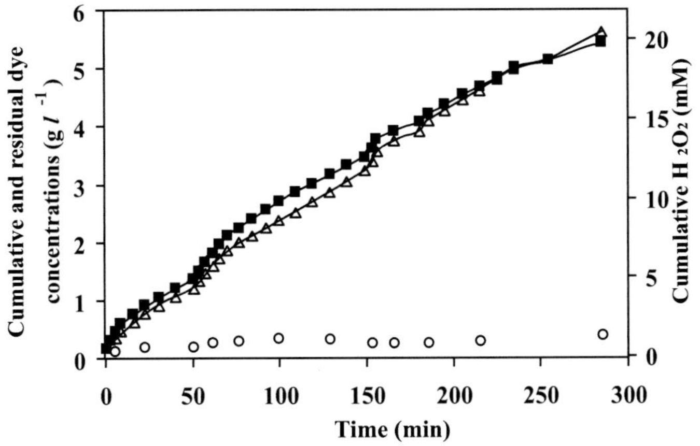

Fig. (36). The cumulative concentration of RBBR dye (■), H_2O_2 concentration (Δ), and residual dye concentration (○) in repeated-batch culture [160].

4.3. Solid-state Cultivation of Recombinant *A. oryzae*

4.3.1. Methods

4.3.1.1. Medium and Strain used

The details of media used in Solid-state Cultivation (SSC) have been described previously [67]. CM1 medium contained glucose, KH_2PO_4, Polypepton, $NaNO_3$, $MgSO_4 \cdot 7H_2O$, and agar. CM2 medium comprised glucose, KH_2PO_4, Polypepton S, $NaNO_3$, $MgSO_4 \cdot 7H_2O$, and agar. These two media were prepared for the pre-cultures and the main culture of transformant *A. oryzae*, respectively. YM medium consisting of yeast extract, maltose, $NaNO_3$, K_2HPO_4, KCl, $MgSO_4 \cdot 7H_2O$, and $FeSO_4$, was used for the liquid culture. Wheat bran powder was added to the CM2 medium as the SSC medium. The composition of wheat bran powder is shown in Table **16**.

A transformant *A. oryzae* holding *dyp* gene was used. The detail of this strain has been described in a previous chapter.

4.3.1.2. Solid-state Culture (SSC)

The suspension of recombinant *A. oryzae* grown in the CM1 medium was used as a seed for the CM2 medium containing wheat bran powder. The initial water contents were set at 50 or 67% after wheat bran powder was supplemented as a solid substrate. The seed was inoculated and the temperature was statically maintained at 27, 30, or 33°C. Humidity was set at 60%.

4.3.1.3. rDyP Activity Analysis

The supernatants of SSC samples were used for rDyP activity and total protein analyses by the methods reported previously [143].

4.3.1.4. Liquid Culture

The recombinant *A. oryzae* cells grown in the CM2 medium were put into the YM medium, and the cell suspension was prepared. A part of the suspension was inoculated into the YM medium and cultivated. The measurement of DyP activity, total protein, and pH was conducted according to the same methods described above.

4.3.2. Results

During the reaction, humidity has been found to be an important factor in the production of rDyP.

When humidity was increased to 75%, the maximum rDyP activity was not stable even at the initial water content of 50%, as shown in Fig. (**37**). The change in rDyP activity at 75% humidity and 60% humidity was similar, indicating that water evaporation from the medium was influenced by high humidity, and this has been found critical for high rDyP activity.

Fig. (**38**) shows the changes in rDyP activity in SSC at a humidity level of 60% under different initial water contents. At the initial water content of 50%, the maximum activity of rDyP was 1190 U/g, and this rDyP activity was maintained until the pH reached 7. At a water content of 67%, the maximum rDyP activity was 877 U/g. However, complete disappearance of the activity was observed within 24 h. As rDyP activity at 67% water content appeared about 1 day earlier than that at 50% water content, the water content of 67% was deemed appropriate for the growth and rDyP activity of the fungus. However, the rapid growth caused a relatively faster rise of pH to more than 7 and inactivation of rDyP. At 50% water content, pH change was advantageous for rDyP production due to a mild increase in pH.

The pH increase and rDyP inactivation were found to be closely related at different temperatures. Fig. (**38**) shows a change in rDyP activity in SSC at 27, 30, and 33°C. At 27°C, the activity of rDyP was the largest of 1750 U/g before pH reached 7, and after 96 h. At 33°C, the lowest rDyP activity and the fastest inactivation of rDyP were observed. Therefore, the optimum conditions for rDyP production in the SSC using wheat bran powder were found to be 60% humidity, 50% water content, and 27°C temperature.

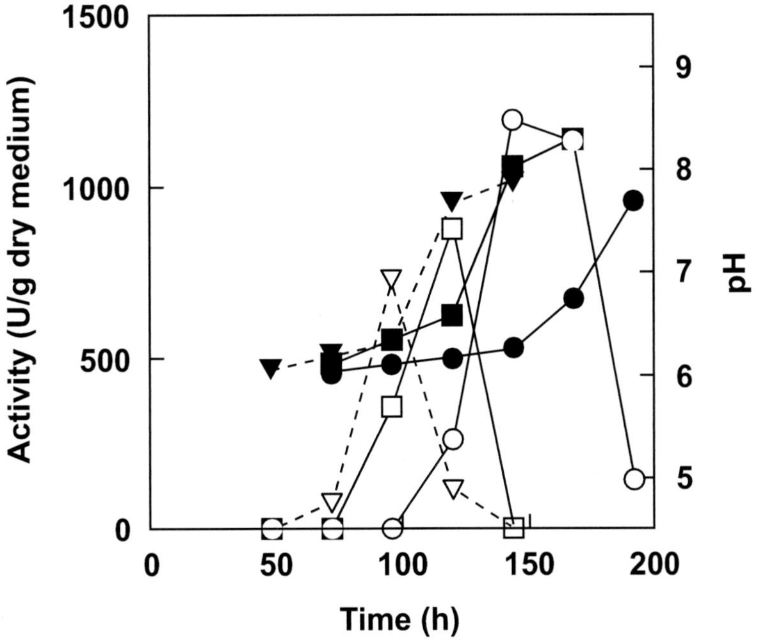

Fig. (37). Changes in rDyP activity and pH in the SSC using wheat bran medium at 50% and 67% water content. Open and closed symbols represent rDyP activity and pH at 50% water content (○ ●) and 67% water content (□ ■) at a humidity level of 60%, respectively. At 75% humidity and 50% water content, changes in rDyP activity and pH are shown by open and closed reverse triangles, respectively [67].

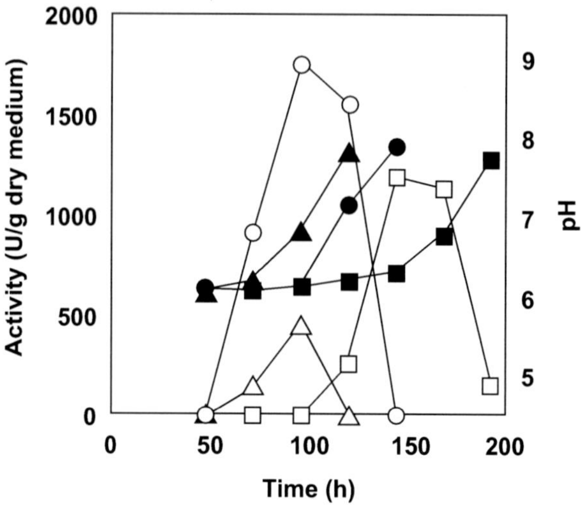

Fig. (38). Changes in rDyP activity and pH in the SSC at 27°C, 30°C, and 33°C at the water content of 50% and humidity of 60%. Open and closed symbols represent rDyP activity and pH, respectively. 27°C (○ ●), 30°C (□ ■), and 33°C (Δ▲), respectively [67].

Generally, enzyme production in SSC is inefficient, especially in large-scale reactors, because the control of several culture conditions is difficult [167]. rDyP production, however, may be appropriate because the critical parameters of the culture control are only the initial water content and the initial humidity. Furthermore, by using an online control system of humidity commercially available for SSC [168], SSC can be one of the alternatives for efficient production of rDyP.

There are a few studies on the comparison between SSC and liquid culture. Some of them are as follows: a comparison of glucoamylase productivity [169 - 173] and a fusion protein in heterologous protein production in *A. oryzae* [174]. Chymosin production in SSC provided approximately 500-fold higher yield than the liquid culture in *A. oryzae* [48].

However, a direct comparison of productivity between SSC and liquid culture has often been controversial. In this work, a new factor, namely, rDyP productivity per gram of carbon and nitrogen in each medium was introduced, and the comparison between the two cultures was conducted. For the comparison, the elemental analysis of the used materials in the two methods was conducted. As a result, the carbon and nitrogen contents of dried wheat bran powder were 44.5% and 3.0%, respectively. On the other hand, the carbon and nitrogen contents of dried yeast extract were 41.0% and 11.1%, respectively. The maximum activity of rDyP in a liquid culture was 31.5 U/g of the medium. From these data, a comparison between the SSC and the liquid culture was carried out, and the results are presented in Table **20**. It is clear that rDyP activity in the SSC was higher than that in the liquid culture in all aspects. As the net amounts of carbon and nitrogen sources per gram medium of the SSC were higher than those of the liquid medium, cell density per gram medium of the SSC was also higher than that of the liquid culture. Therefore, both rDyP and total protein productivity of the SSC were found to be significantly higher than those of the liquid culture. This indicates that enhanced rDyP productivity in the SSC may not be due to the improved rDyP expression efficiency, but due to the increase in total cell mass per gram medium in the SSC.

Table 20. Comparison of rDyP activity and productivity between SSC and liquid culture (the medium was wet gram medium; wheat bran contained 50% water) [67].

		Liquid (A)	Solid-state (B)	B/A
rDyP activity	Unit per gram medium	31.5	875	27.8
	Unit per gram carbon	961	3930	4.09
	Unit per gram nitrogen	22700	59300	2.61
rDyP productivity	Unit per day gram medium	5.25	219	41.7

The value of 331 U/mg of the specific activity of the purified rDyP [143] was equivalent to 5.3 g of rDyP per kg of wheat bran in the SSC. If the same amount of rDyP in the liquid culture was aimed, the necessary weight in the liquid culture was kept at 56 kg of liquid medium. As the rDyP productivity in the SSC was similar to the productivity of other enzymes [48], the SSC can be considered one of the effective and promising methods for producing rDyP.

4.4. Air Membrane Surface (AMS) Reactor Culture of Dec 1

4.4.1. Methods

4.4.1.1. Liquid Culture

The liquid culture method for Dec 1 has been previously reported [1]. The brief explanation of the culture is described as follows:

Dec 1 conidial suspension that was grown onto the PD agar plate was dispersed in distilled water. Then, the suspension was filtered and the filtrate was incubated in MPY medium in a shaking flask as shown in Fig. (**39**) for 5 days at 30°C. The cell-free culture solution was used for the analysis of secreted protein and activity of DyP. The Dec 1 cell mass was expressed as the dry cell mass. MPY medium composition has been shown in a previous chapter.

4.4.1.2. AMS Reactor

The AMS reactor was designed by referring to a previously reported paper [56]. Fig. (**39**) shows a schematic reactor system for the liquid and AMS reactor cultivations. The brief procedure for the handling of the AMS reactor is provided as follows:

A small Petri dish being 35 mm in diameter and 10 mm deep was placed inside a larger glass Petri dish being 90 mm in diameter and 20 mm deep. Sterilized MPY medium was poured into the small dish and a cellulose ester membrane filter 47 mm in diameter and of 20 µm pore size was placed onto the surface of the small Petri dish. Then, Dec 1 conidial solution was poured onto the center of the membrane and static Dec 1 cultivation on the surface of the membrane was performed at 20, 25, and 30°C and 70% relative humidity. The MPY liquid medium beneath the membrane was used for protein and DyP analyses. A portion of grown Dec 1 cells was removed from the membrane periodically, and dried at 100°C.

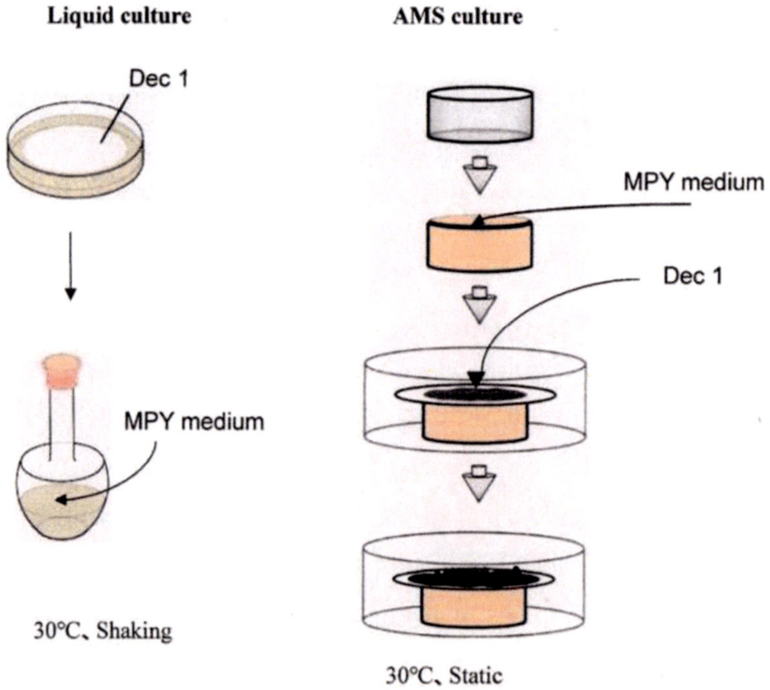

Fig. (39). A schematic representation of the reactor system of liquid and AMS cultivation (prepared by the author).

4.4.1.3. Protein and Enzyme Analysis

Protein and enzyme assays of DyP, MnP, and AAO were followed by the methods described in the previous chapter [22, 23].

4.4.1.4. Dye Decolorization

The dye decolorization method is defined as follows:

Sixteen synthetic dyes used are shown in Table **21**. Each MPY medium sample containing each of the 16 synthetic dyes was cultivated and the decolorization rate of each dye was calculated based on the absorption maxima of each dye as follows:

Table 21. Decolorization rates of various dyes measured by Dec 1 grown in AMS [175].

Dyes	Chromophore[a]	λ max (nm)	Final Concentration (mg/*l*)	Decolorization Rate (%)		
				day 6	day 8	day 10
R. blue 114	AQ	620	200	83.6	99.1	100
R. blue 5	AQ	600	180	66.9	90.1	98.7
R. blue 19	AQ	590	100	69.3	91.1	100
R. blue 182	Azo	610	400	73.1	98.7	100
R. red 202	Azo	548	150	64.9	92.3	98.8
R. red 187	Azo	533	100	46.9	100	100
R. red 123	Azo	500	100	31.9	91.7	100
R. red 225	Azo	530	60	60.6	84.3	100
R. red 120	Azo	535	100	82.9	93.7	92.9
R. orange 13	Azo	490	100	46.5	54.9	75.2
R. yellow 2	Azo	390	200	23.1	38.2	44.4
R. black 5	Azo	598	300	79.7	93.7	99.9
Acid red 73	Azo	507	100	83.1	97.7	100
R. blue 21	PC	665	300	81.7	93.7	87.2
Acid red 27	Azo	526	100	ND[b]	ND	97.9
Acid orange 7	Azo	484	100	ND	ND	100

a: AQ, anthraquinone dye; Azo, azoic dye; PC, phthalocyanine dye.
b: No data available.

(Initial absorption value of a dye minus absorption value

after culture) divided by (initial absorption value) × 100

As a reference, MPY liquid culture was conducted without dyes.

DyP and MnP activities were measured under different water contents. MPY agar plates containing different concentrations of glycerol (v/v) were prepared to decrease water contents and Dec 1 was cultivated on each agar plate [176]. Enzyme activities recovered from each agar plate were measured by the method reported previously [177]. The enzyme activities were converted to the unit (U) per kg of MPY agar medium.

Observation by scanning electron microscopy

The detail of the procedure has been provided previously [175].

4.4.2. Results

4.4.2.1. Secreted Protein and DyP Production in Liquid Culture

In liquid culture for 5 days at 30°C, the results of the maximum dry cell mass, the amount of secreted protein, DyP activity, and AAO activity were 3.62 g/l,1.81 mg/dry cell mass, 42.4 U/l, and 2.01 U/l, respectively. MnP activity was not detected. These data were compared with those obtained from AMS cultures.

4.4.2.2. Secreted Protein and DyP Production by AMS Culture

Figs. (**40A** and **40B**) show changes in dry cell mass of Dec 1 and the amount of secreted protein per mg per dry cell mass at 20, 25, and 30°C, respectively. A gradual decrease in dry cell mass was observed after 16 days at 25 and 30°C presumably due to autolytic degradation of Dec 1. The maximum amount of secreted protein at 20°C, 25°C, and 30°C was found to be 62.5, 32.5, and 14.6 mg/g dry cell mass, respectively. These values were 34.5-, 18.0-, and 8.1-fold higher than those obtained from the liquid culture. Figs. (**41A**, **41B** and **41C**) show changes in DyP, MnP, and AAO activities at 20°C, 25°C and 30°C, respectively. The maximum activities of DyP and MnP at 25°C and of AAO at 30°C were 9.86 U/ml, 5.07 U/ml, and 217 U/l, respectively.

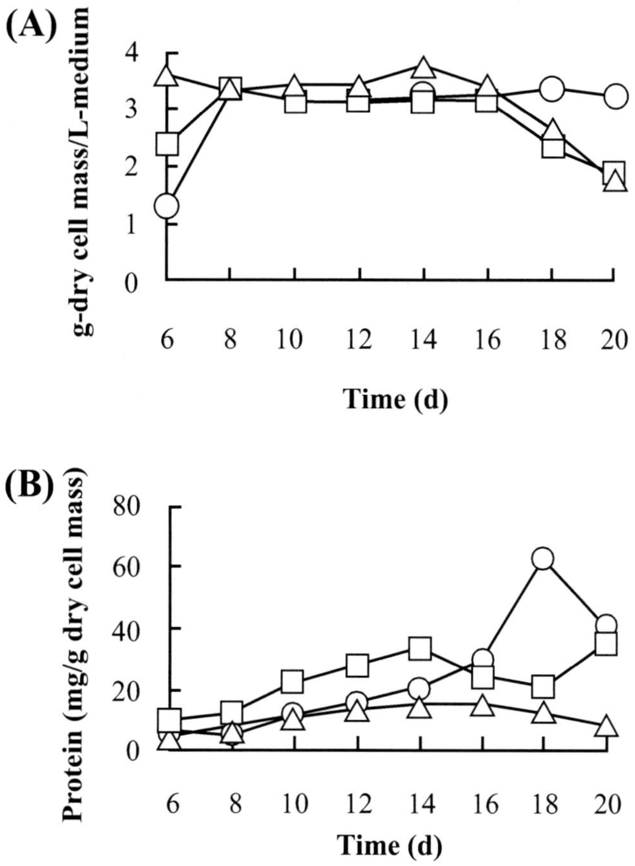

Fig. (40). Changes in dry cell mass of Dec 1 (**A**) and secreted protein (**B**) under different temperature conditions. ○, 20°C; □, 25°C; Δ, 30°C. [175].

4.4.2.3. Decolorization of Dyes by Employing AMS Culture

Table **21** demonstrates the periodically measured decolorization rates for 16 synthetic dyes in AMS culture. All anthraquinone dyes were decolorized at more than 90% of the decolorization rate within 8 days. The decolorization rates of azo dyes, R. orange 13, R. yellow 2, and R. blue 21 were slightly lower than those of other dyes. A similar tendency was also observed in liquid cultures of Dec 1, as detailed in Chapter 2 [1].

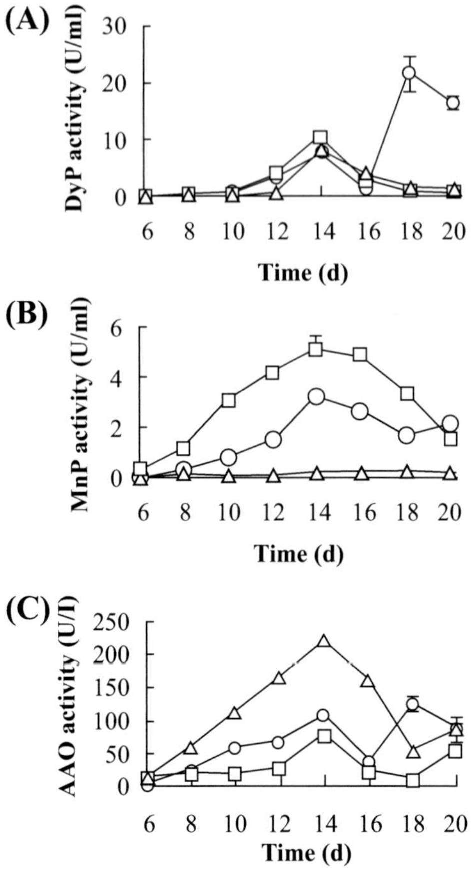

Fig. (41). Changes in DyP (**A**), MnP (**B**), and AAO (**C**) activities under different temperature conditions. ○, 20°C; □, 25°C; Δ, 30°C [175].

4.4.2.4. Effects of Water Content on Enzyme Activities

Figs. (**42A** and **42B**) show changes in DyP and MnP activities under different water contents using different glycerol concentrations, respectively. The time to reach the maximum activity of DyP was delayed with increasing glycerol concentration. However, it was difficult to find any relation between the maximum value of DyP activity and glycerol concentration. On the other hand, maximum MnP activity increased from 520 to 1370 U/kg according to an increase in glycerol concentration in the medium. However, at 10% glycerol concentration, the maximum activity of MnP was delayed by 2 days, and the value was lower than those at other glycerol concentrations. The rate of decline after reaching the maximum activity was the slowest at 10% glycerol concentration.

4.4.2.5. Scanning Electron Micrographs of Mycelia

Fig. (**43**) shows the results of scanning electron micrographs of Dec 1 cells that were grown under different culture conditions. The surface of the mycelium of Dec 1 in the liquid culture sample was smooth and no biofilm was formed, as shown in Fig. (**43A**). On the other hand, Fig. (**43B**) shows that fungal mycelia grown in AMS culture formed biofilm by producing thick viscous material. At glycerol concentrations of 0% and 5%, the surfaces of mycelia were similar to the surface of the liquid culture of Dec 1, as shown in Figs. (**43C** and **43D**). At 10% glycerol concentration, a biofilm was developed, as shown in Fig. (**43E**).

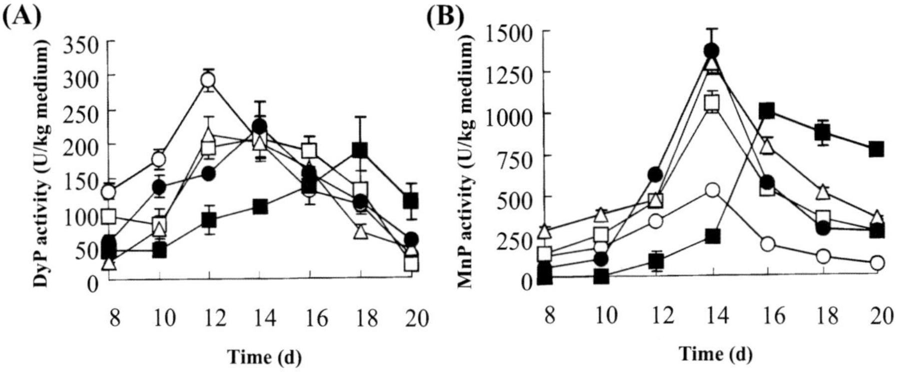

Fig. (42). Change in DyP and MnP activities under different water activities. 0% (○), 2.5% (□), 5% (Δ), 7.5% (●), and 10% (v/v) (■) of glycerol concentration. (**A**) DyP activity; (**B**) MnP activity [175].

4.4.2.6. Enzyme Production by AMS and Liquid Cultures

Table **22** shows the results of maximum enzyme productions by AMS and liquid cultures. The activities of DyP and AAO in AMS culture were 232- and 108-fold higher than those from liquid culture, respectively. MnP activity was not detected in liquid culture, but the apparent activity was observed in AMS culture.

Table 22. Maximum enzyme production by AMS and liquid cultures [160].

	Liquid	AMS	AMS/Liquid
DyP (U/l)	42.4	7770	232
MnP (U/l)	0	5070	
AAO (U/l)	2	217	108

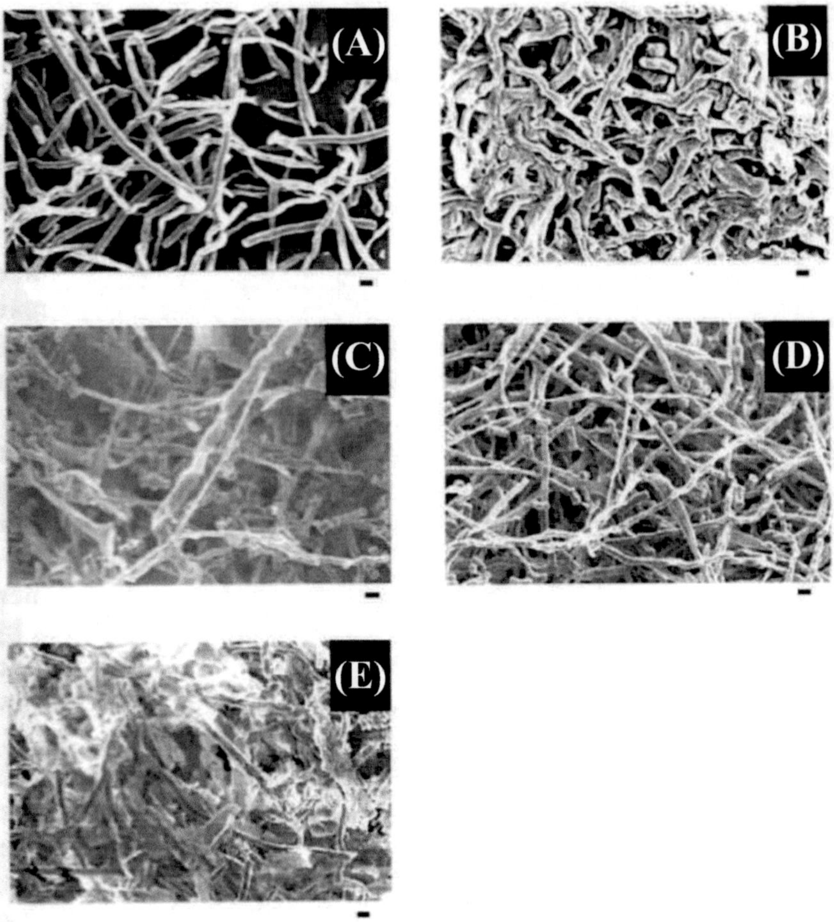

Fig. (43). Results of scanning electron micrographs of Dec 1 under different culture conditions. (**A**) Liquid culture; (**B**) AMS culture; (**C**) MPY agar plate culture without glycerol; (**D**) MPY agar plate culture with 5% (v/v) glycerol; (**E**) MPY agar plate culture with 10% (v/v) of glycerol concentration [175].

4.4.3. Discussion

In the AMS reactor, Dec 1 cells grown on a membrane surface utilized liquid nutrients diffusing through the membrane. An advantage of AMS culture is that protein metabolites produced from the cell mass on the membrane are separated effectively because most of the secreted proteins penetrate the aqueous phase.

AMS culture involves static cultivation, and no mechanical energy is required for agitation, as well as there is no need to remove suspended cells from the liquid phase because the cells and aqueous phases are completely separated by a membrane. This system may be applicable for *in vivo* wastewater treatment if proper continuous flow system and control of the cell mass are ensured.

The dry cell mass of Dec 1 grown at 30°C was the highest, but the amount of secreted protein was the lowest during day 10 to day 16 of culture, as shown in Fig. (**40**). Protein secretion was best at 25°C among three temperatures mainly because dry cell mass did not decrease at 20°C. This suggests that autolysis, which is often observed after exhaustion of nutrients, did not occur at 20°C even after long incubation of Dec 1. In AMS culture, the apparent optimum temperature was set at 25°C, but the temperature inside the mycelial mat might have been higher than that of the surrounding area because the heat generated by metabolism may be accumulated [178].

The maximum dry cell mass was 3.3 g/l of the medium in AMS culture, and this was almost the same as that of liquid culture at 3.62 g/1 of culture. This indicates that Dec 1 had no serious growth problems even though it was a static culture. The secreted amount of protein in AMS culture was significantly higher than that in liquid culture, as shown in Table **22**. A similar result was reported in the SSC of *A. oryzae*, showing protein production in SSC to be 6.4-fold higher than that in liquid culture [51, 55, 57]. The difference between the SSC and MAS systems is as follows: the proteins secreted in SSC accumulated inside the solid mat because of the aqueous phase. However, in AMS culture, the secreted protein was efficiently delivered from the solid cell mass phase to the aqueous phase beneath the permeating membrane. Therefore, produced protein recovery from the aqueous phase has been found to be more efficient in AMS than in SSC, leading to high productivity of DyP.

The production of DyP and AAO in AMS culture at 25°C was 232- and 108-fold higher, respectively, than from liquid culture, as shown in Table **22**. This means that specific dye decolorization enzyme production in AMS culture was significantly higher than that in the liquid culture. Thus, it is obvious from the results that the growth of Dec 1 was similar between AMS and liquid cultures.

MnP activity was not detected in liquid culture although the activity was apparent in the AMS reactor [175]. However, MnP activity at 30°C in AMS culture was significantly low compared to that at 20°C and 25°C, as shown in Fig. (**41**). One possibility may be that in MPY medium, temperature-sensitive isozymes might be expressed predominantly at 30°C because Dec 1 produced MnP isozymes in a previous study [23]. Similar results with other peroxidases from several basidiomycetes have been presented [128, 179 - 181]. The maximum DyP activity was observed at 20°C, which has been found to be significantly higher than other cultivation methods described, presumably because the expression of different isozymes was induced by nutrient contents in media and metal ions in the AMS reactor. The related results have been presented in the next section.

In a previous chapter, AAO activity and simultaneous production of H_2O_2 were observed before DyP activity appeared in liquid culture [22], indicating that AAO plays a role in supplying H_2O_2 to DyP. A similar phenomenon was observed in AMS, as shown in Fig. (**41**). On day 14, both DyP and AAO showed activity peaks, but a significant amount of AAO was produced before day 14. Second peaks of DyP activity and AAO activity appeared on day 18 at 20°C. From these results, AAO can be considered as associated with the expression and activity of DyP.

Dye decolorization by AMS culture seemed to be different from that of liquid culture. In liquid culture, the dyes initially adsorbed to the cells of Dec 1, but the adsorbed color disappeared in accordance with the growth of Dec 1 culture. However, in AMS, Dec 1 decolorized the dyes, but not directly, rather by secreted enzymes in the liquid phase. Although DyP activity was low at 0.5 U/ml until 10 days in AMS, the amount was theoretically enough to decolorize 0.5 μmol of several dyes for a few minutes, as shown previously [23].

Microscopy observation showed the mycelia of AMS cultures to be heavily packed, while those of liquid cultures were dispersed and the biofilms were not formed. Biofilm formation by *Aspergillus niger* was associated with high production of cellulose-degrading enzymes [182, 183]. In SSC, the lowered water activity of hyphae promoted biofilm formation [184]. In this study, scanning electron microscopy compared the mycelia of Dec 1 from AMS cultures and agar plate cultures of Dec 1. Mycelia from AMS cultures were rich in viscous polysaccharides, leading to biofilm formation. 10% glycerol also developed biofilm formation in spite of a gradual decrease in glycerol concentration. It is not clear why low water activity promoted biofilm formation, but the agar plate containing 10% glycerol showed similar fungal growth in the AMS culture. This suggests that DyP and MnP activities may not be closely correlated with biofilm formation because the activities of these enzymes were not maximum at 10% glycerol.

Actually, MnP activity increased with decreasing water activity, as shown in Fig. (**42B**), but biofilm formation on agar plates interrupted enzyme production. Biofilm formation may inhibit enzyme secretion into the environment. However, in AMS culture, high enzyme productivity was observed under biofilm formation, as shown in Fig. (**43B**). This indicates that other factors than water activity may be involved in enzyme transport from the solid phase to the aqueous phase.

4.5. Production of Two Isozymes of DyP in AMS Culture

4.5.1. Methods

The Air Membrane Surface (AMS) reactor and the culture method used have been described in a previous section. The fungus, Dec 1, has been described in detail in Chapter 2.

As the details of the purification procedure have been provided in a previous article [185], the purification steps have been described here briefly. From the culture broth of Dec 1, the supernatant was applied to several chromatography columns, and each fraction of DyP eluted from columns was collected.

4.5.2. Results

The recombinant DyP, previously named rDyP, has been renamed DyP1 here.

Fig. (**44**) shows the characterization of isozymes. The activities of two isozymes, DyP2 and DyP3, were detected, as shown in Fig. (**44A**), and these were purified. The elution time of DyP2 was slightly shorter than that of DyP3, DyP2, and DyP3 revealed bands of different molecular sizes, as shown in Fig. (**44B**). The molecular masses of DyP2 and DyP3 were estimated to be 58 kDa and 47 kDa, respectively.

Table **23** shows the purification steps of isozymes. The specific activity of DyP3 was found to be 732 U/mg, 2.80 times greater than 261 U/mg of DyP2. The recovered amount of DyP3 was too small to discuss the difference in specific activities in detail. The N-terminal amino acid sequence of DyP2 was AXDTILPLNN with an unidentified residue, X [185]. This sequence has been found to be close to that of the DyP1 sequence of ANDTILPLNNIQG. On the other hand, the DyP3 sequence of NLPPDNIQG lacked four residues from the N-terminus of DyP1 and exhibited 67% identity with DyP1.

If the X residue of DyP2 was asparagine, the sequence of DyP2 has been found identical to DyP1, indicating the existence of an N-linked oligosaccharide. In contrast, as DyP3 had no possible site for an N-linked oligosaccharide in the analyzed sequence, the N-terminal sequence of DyP3 was different from that of DyP1 and DyP2.

The effect of pH and temperature on purified enzymes of DyP2 and DyP3 together with DyP1 was investigated, as shown in Figs. (**44C** and **44D**). DyP1, DyP2, and DyP3 decolorized and oxidized various synthetic dyes and several aromatic substrates by the method reported previously [143, 175].

Table 23. Purification of DyP2 and DyP3 from *T. cucumeris* Dec 1by AMS culture [185].

	Volume (ml)	Protein (mg)	Total Activity (U)	Specific Activity (U/mg)	Yield (%)	Purification (fold)
Supernatant	315	25.1	430	17.1	100	1
Ultrafiltration	90	23.6	416	17.6	96.7	1.03
Butyl-toyopearl						
DyP2	49	6.43	366	56.9	85.1	3.33
DyP3	39	1.44	43.3	30.1	10.1	1.76
Superdex 200						
DyP2	2.0	2.00	193	96.5	449	5.64
DyP3	1.5	0.18	25.9	144	6.02	8.42
Mono Q						
DyP2	2.0	0.19	49.6	261	11.5	15.3
DyP3	1.0	0.0079	5.78	732	1.34	42.8

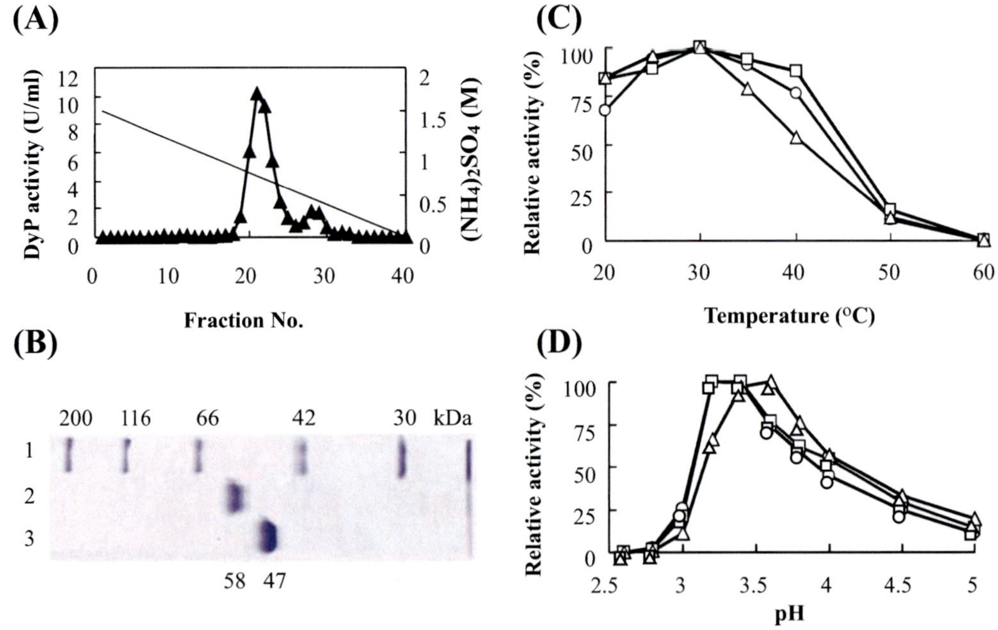

Fig. (44). Characterization of two isozymes, DyP2 and DyP3, produced in AMS reactor. (**A**) Mono Q chromatogram of isozyme isolation. Linear gradient line, ammonium sulfate concentration; ▲, DyP activity. The first peak, DyP2; the second peak, DyP3; (**B**) SDS-PAGE. Lane 1, molecular mass markers; lane 2, DyP2; and lane 3, DyP3. The molecular mass, 58kD of DyP2 and 47kD of DyP3; (**C**) effect of temperature; (**D**) effect of pH on enzyme activities. Symbols, DyP1 (○), DyP2 (□), and DyP3(Δ) [185].

Table **24** presents the substrate specificities at of DyP1,DyP2 and DyP3 relative percentages of the activity of 100% for RB5. DyP2 and DyP1 showed similar substrate similarities to each other, but that of DyP3 was different from the others. The activity of DyP3 was found to be lower than that of DyP2 and DyP1 at 35 °C and 40 °C, as shown in Fig. (**44C**). The optimum pH for DyP3 activity was found to be higher than that of the other enzymes, as shown in Fig. (**44D**) [185]. The substrate specificities of DyP3 for DMP and guaiacol were more than 6.5 times higher than those of DyP1 and DyP2, as shown in Table **24**. All three enzymes showed no activity for veratryl alcohol. Therefore, DyP3 has been judged to be different from DyP1 and DyP2 and considered to be an allelic variant. As DyP2 and DyP1 were found to be similar in various aspects, these two enzymes have been considered to have the same genetic origin and differ slightly in the presence or absence of an oligosaccharide component. At present, increasing numbers of reports on the DyP-type peroxidase family are present, as described in Chapter 1.

Table 24. Relative activity (%) of various substrates toward R. blue 5(100%) [185].

	DyP1	**DyP2**	**DyP3**
R blue 19	94	84	160
R blue 21	7.5	9.6	38
R red 123	0.72	1.4	0.80
R red 202	0	0	0
R red 225	8.2	14	9.9
AQ - 2 [a]	77	85	150
VA [b]	0	0	0
DMP [c]	10	11	65
Guaiacol	5.6	6.6	37

a: 1-amino-2-sulfonyl-4-aminomethyl-9, 10-anthraquinone.
b: veratryl alcohol.
c: 2, 4-dimethoxy-phenol.

4.6. Productivity Enhancement of DyP

The enhancement degree of DyP production by the different cultivation methods attempted in previous sections is summarized in Table **25**. In all cultivation methods, DyP activities were the values under optimized culture conditions. The gene *dyp* introduced by *A. oryzae* culture increased the production of recombinant DyP 400 times more than the original one. The two reactor systems, AMS reactor and solid-state reactor, enhanced the production of the recombinant DyP, rDyP, significantly. Finally, the initial DyP activity was enhanced by about half a million times after AMS culture and recombinant *A. oryzae* showed the same level of enhancement of DyP with Dec 1. This means that direct use of rDyP is possible in the treatment of colored substance-containing wastewater.

Table 25. The enhanced degree of peroxidase, DyP, by different cultivation methods (prepared by the author).

Strain	DyP Activity (U/l)	Cultivation Method
Thanatephorus cucmeris Dec 1 (original strain)	5.5	Liquid culture
Aspergillus oryzae (gene *dyp* introduced strain)	2020	Liquid culture
Aspergillus oryzae	84234(2020x41.7)	Solid-state culture
T. cucumeris	468,640 (2020x232)	AMS reactor

CONCLUSION

Cloning of a novel peroxidase gene, *dyp*, was successful, and the *dyp* gene was transformed into *A. oryzae*. The recombinant DyP, rDyP, produced from recombinant *A. oryzae* was purified and crystallized. X-ray diffraction data of crystallized rDyP revealed DyP to have a unique tertiary structure different from that of most other peroxidases. When the N-terminal amino acid sequences of the native DyP and recombinant rDyP were analyzed, the absence of a histidine residue in both DyPs, important for heme binding of DyP, was confirmed. This indicates that rDyP produced by *A. oryzae* and native DyP exhibited a similar function.

DyP contains four-stranded antiparallel β-sheets and the arrangement on the distal side of the heme pocket of DyP is completely different from the structures of other known peroxidases, like LiP and MnP. From these data, DyP has been judged to be a novel peroxidase, forming an independent family separated from the classical peroxidase groups. We designated this as "DyP-type peroxidase family".

The productivity of rDyP from *A. oryzae* was about 400 times higher than that of native DyP from Dec 1. Attempts have been made to obtain enhanced productivity of rDyP using different cultivation methods. When rDyP production was conducted using recombinant *A. oryzae* by employing the repeated-batch method using a synthetic liquid medium, a significantly higher number of cycles of repeated-batch culture and a high level of rDyP activity were obtained. When cultures were maintained at 4°C for 20 days, rDyP production resumed in a single day of incubation. Crude rDyP activity of 4600 U decolorized 5.07 g anthraquinone dye, RBBR, at a high decolorization rate of 17.7 mg /l/min.

When the productivity of rDyP by recombinant *A. oryzae* was compared in liquid culture using complex substrates in both repeated-batch and fed-batch cultures, all average rDyP productivities obtained by these cultures were significantly higher than those for any other peroxidases previously reported in liquid cultivations.

rDyP production was also conducted using Solid-state Culture (SSC). Under the optimum conditions for rDyP production in the SSC, the productivity of rDyP was 5.3 g per kilogram as a solid medium. This value of 5.3 g per kg medium corresponded to the productivity of a 56 kg liquid culture. A comparison between SSC and liquid culture has been conducted on different nutritional bases, and DyP productivity in the SSC has been obtained as higher than that in the liquid culture in all compared bases.

To solve the demerits of liquid culture and SSC and utilize the advantages of these cultures, an Air Membrane Surface (AMS) reactor culture, which is a kind of hybrid reactor between solid-state and liquid cultures, has been introduced for Dec 1 growth. Although the optimum temperatures for AMS and liquid culture were different, Dec 1 growth using the two methods was almost the same. DyP production, DyP activity, and Aryl Alcohol Oxidase (AAO) activity were significantly higher in the AMS culture than in the liquid culture, showing the advantage of AMS culture over other cultivation methods. Two DyP isozymes, DyP2 and DyP3, were also purified from the culture broth of Dec 1 in the AMS reactor and characterized.

CHAPTER 5

Dye Decolorization by Immobilized Recombinant rDyP and Turnover Capacity of rDyP

Abstract: For an efficient application of the soluble recombinant enzyme of rDyP, immobilization of the enzyme was caried out to enhance and stabilize catalytic efficiency of rDyP. Although several conventional immobilization methods have been attempted for rDyP, no methods have been successful. Therefore, new catalysts developed for exhaust gas removal were employed. They were silica-based mesocellular foams and two silica-based porous materials, FSM-16 and AISBA-15, which were chemically synthesized. Immobilization of rDyP on them was carried out and immobilization efficiency was assessed. The overall efficiency was defined as adsorption efficiency x activity efficiency to find the maximum efficiency. The efficiency of rDyP immobilized on FSM-16 and AISBA-15 was maximum at pH 5 and pH 4, respectively. FSM-16 showed advantages over AISBA-15 in terms of stronger affinity for rDyP due to its anionic surface and much lower leaching of rDyP from FSM-16. When the rDyP immobilized on FSM-16, an anthraquinone dye, RBBR, was decolorized in repeated-batch mode, and eight sequential batches were possible, while rDyP immobilized on AISBA-15 enabled only two batches.

For evaluation of the practical potential of rDyP, the turnover capacity of rDyP was introduced. In order to minimize H_2O_2 inactivation for rDyP activity, four H_2O_2 supply methods were attempted and the turnover capacity of each method was compared. The continuous fed-batch supply of H_2O_2 and the stepwise fed-batch supply of the dye gave the maximum turnover capacity of 20.4. At this turnover capacity, one liter of crude rDyP solution containing 5,000 U could decolorize up to 102 g dye in 10h.

Keywords: AISBA-15, Adsorption, FSM-16, Immobilization, Inactivation by H_2O_2, Leaching, Mesoporous materials, Overall efficiency, Remazol Brilliant Blue R, Recombinant DyP, rDyP, Turnover capacity.

1. IMMOBILIZATION OF RDYP ON MESOPOROUS MATERIALS

1.1. Properties of Newly Synthesized Mesoporous Materials

The synthesis and characteristics of newly synthesized mesoporous materials of FSM-16 and AISBA-15 have been described in detail previously [186, 187]. Table **26** presents the structural and chemical properties of two synthesized materials, FSM-16 and AlSBA-15.

Makoto Shoda
All rights reserved-© 2024 Bentham Science Publishers

Table 26. Structural and chemical properties of FSM-16 and AISBA-15 [186].

	Pore Volume (cm^3/g)	Pore Diameter (nm)	Specific Surface Area (m^2/g)	External Specific Surface Area (m^2/g)	Si/Al
FSM-16	1.9	5.4	1108	66	-
AISBA-15	1.29	10.6	918	22	30

-: not measured., Si/Al: silicate/aluminum

1.2. Effect of pH on Adsorption and Activity Efficiency of rDyP Immobilized on the Mesoporous Materials

The adsorption efficiency is defined as [(Ai – Ar)/Ai] x 100

Where, Ai is the initial rDyP activity before immobilization and Ar is the last rDyP activity after immobilization.

The expected activity of rDyP is defined as (Ai – Ar)/concentration of the material.

The activity efficiency is defined as (measured rDyP activity/expected rDyP activity)x100. Thus, the overall efficiency that reflects the overall activity performance of the immobilized rDyP is defined as adsorption efficiency x activity efficiency.

The immobilization procedure have been described previously [186]. rDyP was immobilized on the mesoporous materials under different pHs. Table **27** shows the result of FSM-16. The adsorption efficiency of rDyP on FMS-16 increased when pH declined from 6 to 3, and expected efficiencies were higher at pHs 3 and 4. Table **28** shows the result of MISBA-15. Although adsorption efficiency was 98% when rDyP was immobilized at pH 3, it was only 20% at pH 6. From the two tables, it is obvious that the expected efficiency of rDyP on FMS-16 was significantly higher than that on AISBA-15. However, the activity efficiency of rDyP on AISBA-15 was higher than that on FSM-16 at pH 4, 5, and 6. Therefore, in order to evaluate the overall efficiency, adsorption efficiency x activity efficiency was defined. As a result, the maximum values of the overall efficiency were 28% at pH 5 for FSM-16 and 43% at pH 4 for AISBA-15. The actual measured activity of rDyP was the maximum, being 2800 U/g-FSM-16 at pH 5, as shown in Table **27**.

Table 27. Characterization of rDyP immobilization on 1 g FSM-16 under different pH conditions (initial rDyP activity used = 10,000U) [186].

pH	Adsorption Efficiency (%)	Expected Activity (U/g-FSM-16)	Measured Activity (U/g-FSM-16)	Activity Efficiency (%)	Overall Efficiency (%)
3	98	9800	933	9.5	9.3
4	97	9700	1833	19	18
5	83	8330	2800	34	28
6	66	6600	2100	32	21

Table 28. Characterization of rDyP immobilized on 1 g AISBA-15 under different pH conditions at the initial rDyP activity of 10,000U [186].

pH	Adsorption Efficiency (%)	Expected Activity (U/g-AISBA-16)	Measured Activity (U/g-AISBA-16)	Activity Efficiency (%)	Overall Efficiency (%)
3	98	9830	833	8.5	8.3
4	97	4360	1933	44	43
5	17	1660	1600	96	16
6	20	1960	1433	73	15

1.3. Leaching of Immobilized rDyP from Mesoporous Materials

As leaching of immobilized rDyP from mesoporous materials reduces the activity of immobilized rDyP, the leaching degree of immobilized rDyP was evaluated by washing the immobilized rDyP repeatedly in buffer solution. Fig. (**45A**) shows the change in rDyP activity immobilized on FSM-16 after leaching under different pH conditions. The rDyP activity decreased by more than 57% at pHs 5 and 6. However, no changes in activity of rDyP were observed at pHs 3 and 4. Fig. (**45B**) shows the result of AISBA-15. The decrease in the activity of rDyP on AISBA-15 was more significant compared to that on FSM-16. No change in activity of rDyP on FSM\16 was observed at pHs 3 and 4.

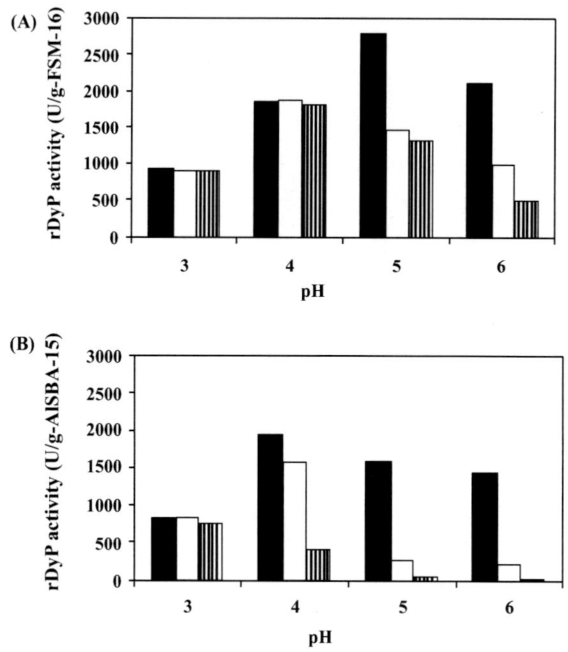

Fig. (45). Effect of leaching on rDyP activity immobilized on FSM-16 (**A**) and on AISBA-15 (**B**) under different pH conditions. Symbols: black bars indicate results on day 1, white bars on day 5, and dark verticals on day 10 [186].

1.4. Stability of Free and Immobilized rDyP When Exposed to H_2O_2

Fig. (**46**) shows the change in the stability of free rDyP and rDyP immobilized on FMS-16 at pH 3. When rDyP was exposed to H_2O_2 in the absence of the dye, it was clear that the stability of immobilized rDyP against H_2O_2 was higher than that of free rDyP. The stability of rDyP immobilized on AlSBA-15 at pH 3 was also higher than that of free rDyP (data not shown).

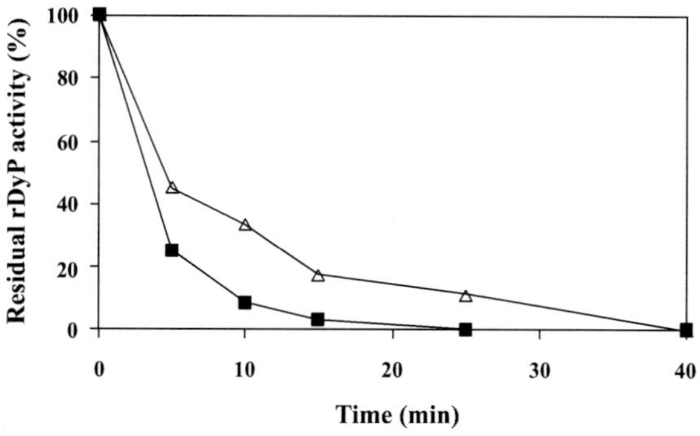

Fig. (46). Change in the stability of free rDyP (■)and rDyP immobilized on FSM-16 (Δ) when exposed to H_2O_2 at pH 3 (initial rDyP activity: 400 U/1) [186].

1.5. Effect of pH on Stability of Free rDyP

Fig. (**47**) represents the stability of free rDyP activity at different pH values. The initial rDyP activity remained stable at more than 90% for 48 h, showing the adverse pH effect on rDyP to be negligible. This stable rDyP activity under different pHs may be associated with high adsorption efficiency of rDyP at pH 3 and 4, as shown in Tables **27** and **28**. Therefore, the decline in rDyP activity, shown in Fig. (**46**), may be mainly because of rDyP inactivation by H_2O_2 and not due to the effect of pH.

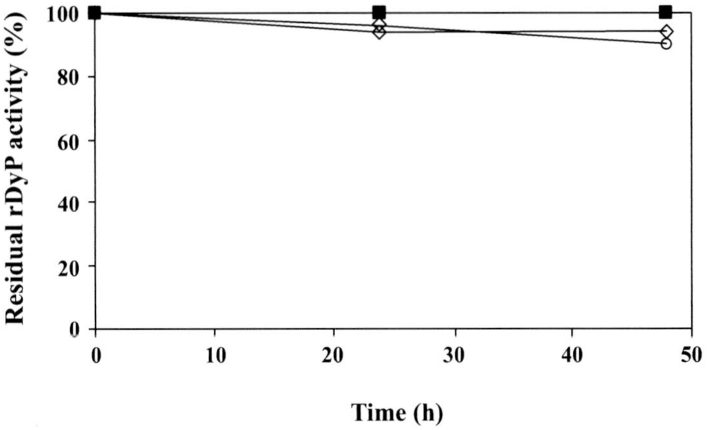

Fig. (47). Stability of free rDyP activity at different pH values. Symbols: pH 3 (○), pH 4 (■), and pH 5 (◇) (initial rDyP activity 1000U/1) [186].

1.6. Decolorization of Dye by rDyP Immobilized on Mesoporous Materials in Repeated-batch Mode

Fig. (**48**) shows the change in decolorization of the dye, RBBR, by rDyP immobilized on FSM-16 under different pH values in repeated-batch mode. Fig. (**48A**) demonstrates that RBBR was steadily decolorized in three batches at pH 3, but a significant decrease in decolorization rate was observed in the fourth batch. Fig. (**48B**) shows that at pH 4, stable decolorization lasted for seven batches. However, a decrease in decolorization rate was seen after four batches at pH 5, as shown in Fig. (**48C**).

Fig. (**45**) shows that no leaching of immobilized rDyP from FSM-16 occurred at pH 3 or 4, but at pH 5, leaching of rDyP was obvious. The result in Fig. (**48C**) reflects the result presented in Fig. (**45**) and rDyP activity decreased from the first batch to the fourth batch gradually, suggesting that less than 10% of the rDyP was adsorbed on the FSM-16 after the fourth batch.

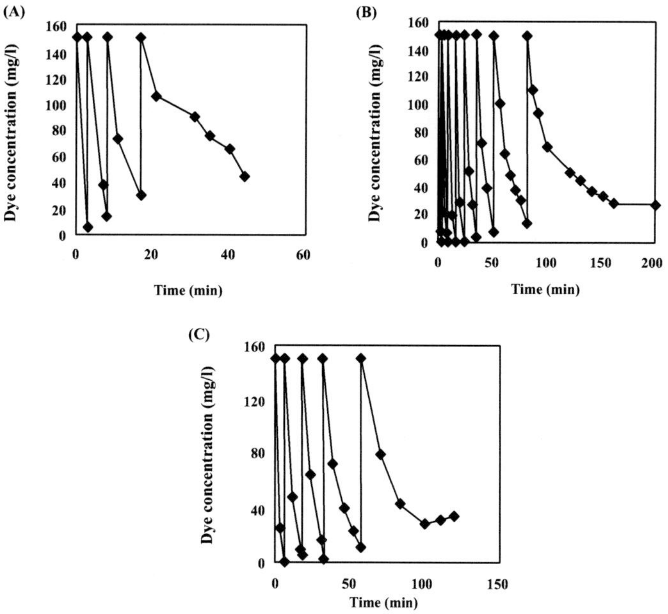

Fig. (48). Decolorization of RBBR by rDyP immobilized on FSM-16 at pH 3 (**A**), pH 4 (**B**), and pH 5 (**C**) in repeated-batch mode [186].

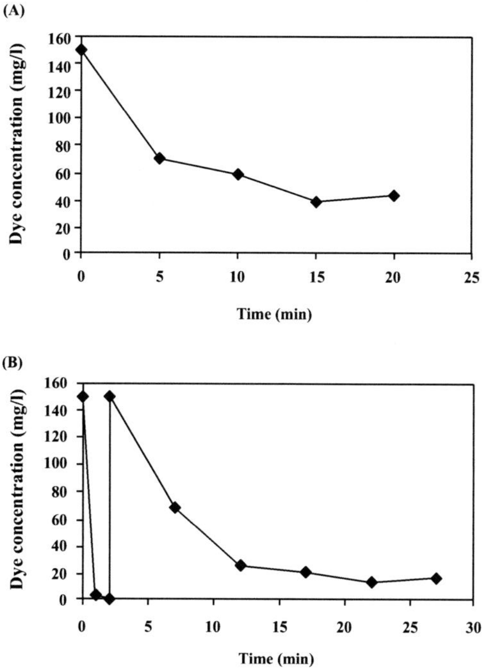

Fig. (49). Decolorization of RBBR by rDyP immobilized on AISBA-15 at pH 3 (**A**) and pH 4 (**B**) in repeated-batch mode [186].

On the other hand, no decolorization was seen by rDyP immobilized on AlSBA-15 at pH 3, as shown in Fig. (**49A**). At pH 4, in only two batches, RBBR decolorization was possible, as shown in Fig. (**49B**). As rDyP in the first batch mode showed only 45% of the activity of rDyP before the experiment, the decreased activity of rDyP immobilized on AIBSA-15 may be mainly due to leaching of rDyP from AIBSA-15.

Another silica-based Mesocellular Foam (MCF) containing cargo-like pores with high porosity was also synthesized and the overall efficiency of 85% after 20 batches of BRRB decolorization with the immobilized rDyP was obtained [186].

1.7. Discussion

Some chemical factors involved in the immobilization of DyP on silica-based mesoporous materials, FSM-16 and AISBA-15, were explored. The surface charges of enzymes and silica materials are important [188-190].

The difference in electric charges between DyP and mesoporous materials has been found to be a primary factor. These silica-based mesoporous materials synthesized here have been negatively charged above pH 2 because the isoelectric point (p*I*) was 2. The isoelectric point of rDyP was previously determined as 4.2 [143]. This means that rDyP has a positive charge below pH 4.2 and a negative charge above pH 4.2. Therefore, pH values were used to determine the electrostatic interaction between rDyP and the mesoporous materials. At the pH range of 2 and 4.2, electrostatic interactions between the two materials were fairly strong, resulting in high adsorption efficiency, as shown in Tables **27** and **28**. On the other hand, above pH 4.2, the surfaces of both materials exhibited negative charges, leading to repulsion between the two materials and resulting in a lower rDyP adsorption efficiency, as shown in Tables **27** and **28**.

At the pH range of 4 and 6, FSM-16 and AISB-15 demonstrated different surface characteristics. This has been reflected in that rDyP adsorption efficiency was higher for FSM-16, as shown in Table **27**, than that for AlSBA-15, as shown in Table **28**. The cationic surfactant of CTMA-Cl was used in FSM-16 synthesis, but the non-ionic surfactant of Pluronic 123 was used in AlSBA-15 synthesis [64, 186]. Therefore, the anionic surface of FSM-16 had a promoting effect on the adsorption of rDyP of positively charged amino acid groups.

The other factor evaluated was the pore size of mesoporous materials and the molecule size of rDyP [64]. The adsorption efficiency of rDyP on FSM-16 was relatively high. There may be two possibilities for this result. Firstly, the actual pore size of FSM-16 was estimated to be approximately 6.85 nm, which was found to be bigger than the pore size of 5.4 nm calculated by the BET method

[186]. This value has been found to be large enough for rDyP to remain inside FMS-16, if FSM-16 keeps an ideal 2-dimension hexagonal channel. Secondly, a possibility can be that adsorption of rDyP occurs at the pore mouth and/or on the external surface of FSM-16. High adsorption of rDyP on FMS-16 was suggested to be both due to the decrease in pH and the stronger electrostatic interaction between them. Actually, the activity efficiency of immobilized rDyP was significantly low. If rDyP conformation was changed by the strong electrostatic interaction of rDyP with the mesoporous material surface, the improper orientation of active sites of rDyP may have partly occurred in the reactions because of decreasing pH. In separate experiments, when rDyP immobilized on mesoporous materials in a buffer solution of pH 3 was transferred into a buffer solution of pH 7, the immobilized rDyP was released from the materials and a free active form of rDyP was recovered. This observation may support the speculation described above. Changes in the activity of immobilized enzymes under different pH conditions were also reported for Chloroperoxidase (CPO) [191] and lipase [192].

Leaching of rDyP from the two mesoporous materials decreased the activity of rDyP at pH values higher than 5 and 4, as shown in Fig. (**45**). As the activity of immobilized rDyP was lost when the mesoporous materials were washed, the decrease in rDyP activity at pH higher than 5 may have been because of leaching of the rDyP from the mesoporous materials, not because of rDyP denaturation or inactivation.

How to supply H_2O_2 has been found to be important to control the stability of rDyP, as shown in a previous chapter. For example, the half-life of 400 U/1 free rDyP was 3, 9, and 24 min at pH 3, 4, and 5, respectively, at a supply of 0.12 mM H_2O_2. The pH has also been determined to be a key factor for the stability of enzymes. The dependence of enzyme half-life on pH was reported for other peroxidases of CPO and HRP [61, 193], implying that an increase in pH induces an increase in the rDyP half-life. However, in the case of rDyP, a pH change from 3 to 5 increased the leaching of rDyP from the two materials, indicating that an increase in pH has an adverse effect on the number of repeated batches. Therefore, the pH has effects both on leaching and the half-life of DyP.

RBBR decolorization did not take place by free rDyP at pH 3 [62], but RBBR was decolorized by rDyP immobilized on FSM-16 at pH 3, as shown in Fig. (**48A**). Furthermore, the instability of rDyP immobilized on FSM-16 against H_2O_2 was lower than that of free rDyP, as shown in Fig. (**46**). These two properties may be associated with structural changes in the micro-environment of the immobilized rDyP [194, 195]. These new materials may help other enzymes that are difficult to immobilize on conventional immobilization materials and increase their stability upon H_2O_2 activation.

2. CHANGE IN TURNOVER CAPACITY OF RDYP

2.1. Methods

The strain used, the production of the crude recombinant DyP, rDyP, the method of dye decolorization, and the rDyP activity assay have been described in the previous chapters. The anthraquinone dye, Remazol Brilliant Blue R (RBBR) was used. The turanover capacity was defined as the milligram of RBBR decolorized per unit of rDyP inactivated over the catalytic life time of rDyP [62].

2.2. Experimental Conditions

Table **29** shows the details of the experimental conditions for the four experimental objectives of C1~C4 for decolorization of a dye, RBBR [62].

Table 29. Experimental conditions of C1-C4 [62].

Experimental Conditions	Initial rDyP Activity (U/1)	Initial Working Volume of Reactor (ml)	Mode of H_2O_2 Addition	Dye Addition in Batch or Each Fed-batch (mM)	Total Amount of Added H_2O_2 (mM; A)	Total Amount of Added Dye (mM; D)	Total Amount of Treated Dye (mM; E)	Ratio of A/D	Ratio of A/E
C1	200	50	0.24 mM (B)	0.24 (B)	0.24	0.24	0.24	1	1
C2	200	50	0.24 (mM) / 12-30 min (SFB)	0.24-0.45 (SFB)	2.68	1.92	1.84	1.4	1.46
C3	200	250	0.24 (mM) / 12-30 min (SFB)	1.48-1.75 (SFB)	8.4	4.8	4.44	1.8	1.89
C4	180	250	0.0184 (mM/ min; CF)	1.6 (CF and FB)	12	6.4	5.82	1.9	2.06

B, batch; SFB, stepwise fed-batch; CF, continuous fed-batch; FB, fed-batch

A brief explanation of these experiments is as follows:

C1. Batch addition of equimolar H_2O_2 and dye

Equimolar amounts of 0.24 mM H_2O_2 and the dye were used for decolorization using 200 U/1 rDyP. The turnover capacity was calculated based on an initial rDyP activity of 200 U/1.

C2. Stepwise fed-batch addition of H_2O_2 and dye

Decolorization by stepwise fed-batch addition of 0.24 mM H_2O_2 and 0.24~0.45 mM dye was conducted with the initial molar ratio range of H_2O_2 to dye being 0.53~1. The turnover capacity was determined in the same way as that mentioned for C1.

C3. Stepwise fed-batch addition of H_2O_2 with a higher concentration of dye

Decolorization by stepwise fed-batch addition of 0.24 mM H_2O_2 and 1.48~1.75 mM dye was conducted at the initial molar ratio of H_2O_2/dye of 0.14-0.16. The turnover capacity was also determined in the same way as that used for C1.

C4. Fed-batch addition of dye and continuous fed-batch addition of H_2O_2

Continuous fed-batch addition of H_2O_2 at a feeding rate of 0.0184 mM/ min and fed-batch addition of 1.6 mM dye were conducted. The turnover capacity was determined based on an initial rDyP activity of 180 U/1.

2.3. Results

2.3.1. C1. Batch Addition of Equimolar H_2O_2 and Dye

Fig. (**50**) shows the decolorization ratio of the dye, RBBR, and residual rDyP activity in batch mode addition using equimolar H_2O_2 and the dye. The dye decolorization ratio reached 95% in the first 10 min and the decolorization rate during 10 min was 14.3 mg-dye/l/min. After the initial decline of rDyP activity, the stable activity at 85% of the initial activity was maintained until 40 min. The decolorization ratio decreased at 40 min when the color of the reaction mixture changed from clear blue to purple, indicating the formation of new by-products or temporary reverse reactions. Therefore, 0.12mM H_2O_2 was added again to the reaction mixture in 40 min, and the decolorization ratio reached 100%, reflecting that the initial H_2O_2 was utilized completely in 40 min. However, the second addition of H_2O_2 inactivated rDyP, leading to a decline in the residual rDyP activity to 50% of its initial value. The dye-decolorization process by rDyP presented spectra changes of the dye, as shown in Fig. (**51**), confirming that the colored by-product was formed after 10 min (B) and then rDyP degraded it after re-addition of H_2O_2 (C). Therefore, aiming at the full decolorizing activity of rDyP, fed-batch addition or continuous fed-batch addition modes were introduced as conducted in the experiments C2-C4. The turnover capacity was 4.75 mg, as shown in Table **30**.

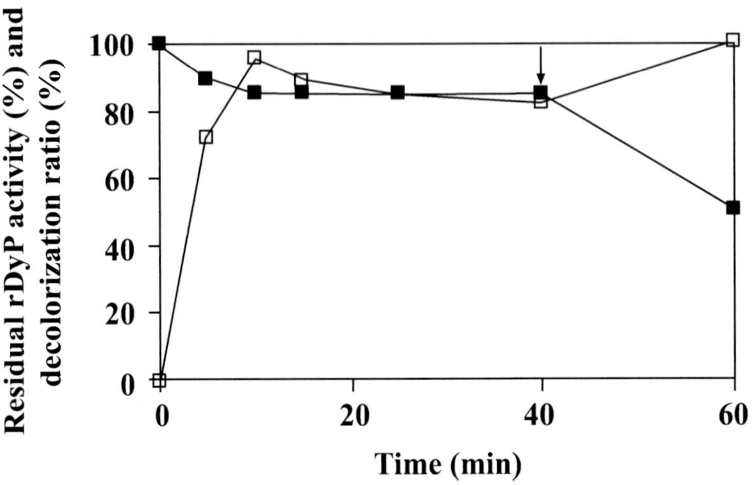

Fig. (50). C1. Decolorization ratio (%) of the dye, BRRB(□), and residual rDyP activity (%)(■) during batch mode treatment of 0.24 mM of H_2O_2 and the dye by the initial 200 U/1 of rDyP. The arrow indicates the time of re-addition of 0.12 mM H_2O_2 [62].

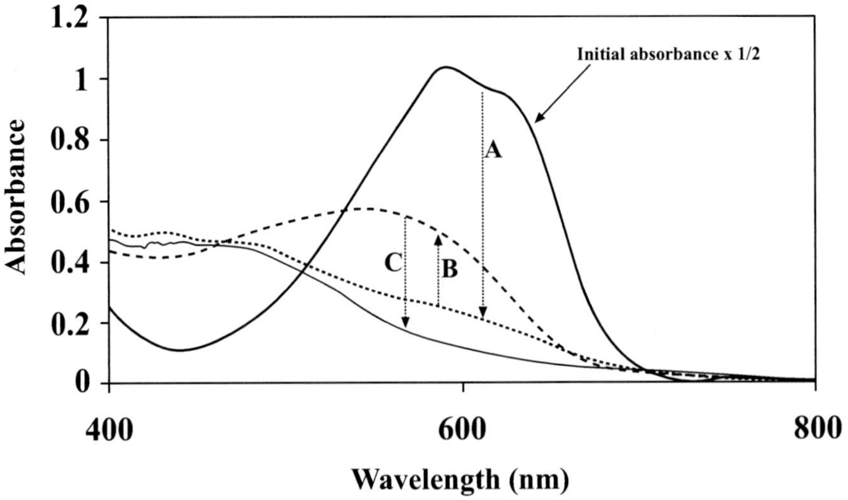

Fig. (51). Spectral changes during the 0.24 mM dye decolorization by 0.24 mM of H_2O_2 in experimental condition C1. **A**, Spectral range decreased in 10 min after the addition of H_2O_2; **B**, spectral range increased from 10 to 40 min of reaction; **C**, spectral range decreased after re-addition of 0.12 mM H_2O_2 at 40 min, as shown in Fig. (**50**) [62].

Table 30. Results of the decolorization of dye, RBBR, by rDyP under different H_2O_2 supplementations (C1~C4) [62].

Experimental Conditions	Inactivated rDyP (U/1; A)	Total Dye Treated (mg/1; B)	Decolorization Ratio (%)	Turnover Capacity (B/A)[a]
C1	30	142.5	95	4.75
C2	200	1,152	87	5.76
C3	200	2,875	92	14.3
C4	180	3,650	92	20.4

a: Turnover capacity is defined as the mg of the dye, RBBR, decolorized per unit of rDyP inactivated.

2.3.2. C2. Stepwise Fed-batch Addition of H_2O_2 and Dye

Fig. (**52**) shows the change in the dye concentration and cumulative H_2O_2 concentration in C2. The 200 U/l of rDyP decolorized 1152 mg-RBBR/l during 4 h. Decolorization rates were calculated as 12.5 mg/1/min in the first fed-batch addition of the dye and 8.7 mg/1/min in the fourth fed-batch addition of the dye, and then the rate decreased to 0.8 mg/1/min at 150 min presumably due to severe inactivation of rDyP by H_2O_2. Then, the fresh crude rDyP of 200 U/1 was added at 240 min. As decolorization resumed, inactivation of the original rDyP by H_2O_2 was confirmed. The average decolorization rate was 7.6 mg-dye/1/min during the whole experimental period. Table **29** shows an A/E of 1.46, *i.e.*, the ratio of total molar H_2O_2 used to total molar dye treated. This indicates that more than a stoichiometric amount of H_2O_2 was required for the complete dye decolorization. The turnover capacity was 5.76, slightly higher than that in C1, as shown in Table **30**.

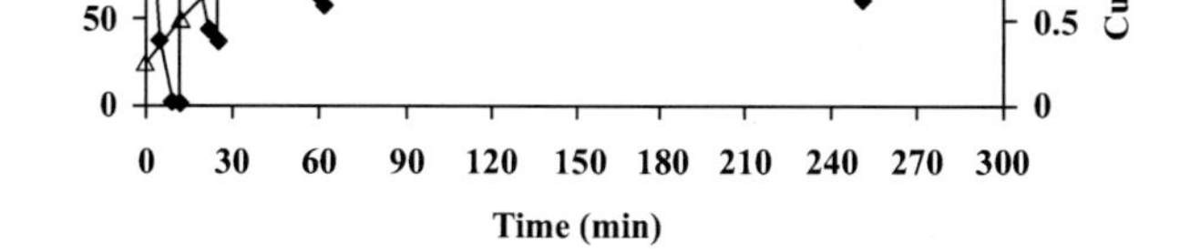

Fig. (52). Change in the dye concentration (♦) and cumulative H_2O_2 concentration (Δ) in stepwise fed-batch additions of the dye and H_2O_2 in the experimental condition C2. The arrow indicates re-addition of 200 U/1 rDyP [62].

2.3.3. C3. Stepwise Fed-batch Addition of H_2O_2 at a Higher Concentration of Dye

Fig. (**53**) shows a change in the dye concentration and cumulative H_2O_2 concentration when H_2O_2 was added in stepwise fed-batch mode and more than 1 mM of the dye was supplied. rDyP of 200 U/1 decolorized 2,875 mg/1 of the dye in 13 h. The decolorization rate of 7.84 mg/1/ min in the first fed-batch of dye addition reduced to 2.1 mg/1/min in the third fed-batch due to the inactivation of rDyP by H_2O_2, and after that, no further decolorization occurred. The average decolorization rate was 5.3 mg/l/min. The ratio of A to E of 1.89, as shown in Table **29**, was similar to that in C2, indicating that more than a stoichiometric amount of H_2O_2 was required for complete decolorization of the dye. The turnover capacity was 14.3, as shown in Table **30**.

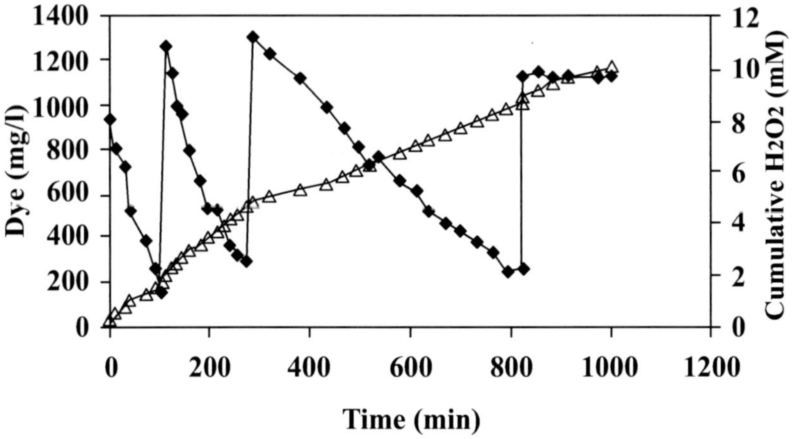

Fig. (53). Change in the dye concentration (♦) and cumulative H_2O_2 concentration (Δ) in stepwise fedbatch addition of H_2O_2 with higher dye concentration in the experimental condition C3 [62].

2.3.4. C4. Fed-batch Addition of Dye and Continuous Fed-batch Addition of H_2O_2

In C4, the continuous fed-batch addition of H_2O_2 and fed-batch addition of a higher concentration of the dye were attempted to minimize rDyP inactivation by H_2O_2. Fig. (**54**) shows a change in the dye concentration and cumulative H_2O_2 concentration. 180 U/1 rDyP decolorized a total of 3,650 mg of the dye/1 in 10 h, and the turnover capacity was 20.4, as shown in Table **30**. By comparing all modes of C1~C4, the lowest rate of inactivation of rDyP was confirmed in C4. As the decolorization rate of 7.6 mg/1/min in the first fed-batch reduced to 4.5 mg/1/min in the fourth fed-batch, partial inactivation of rDyP remained even in continuous fed-batch supply of H_2O_2.

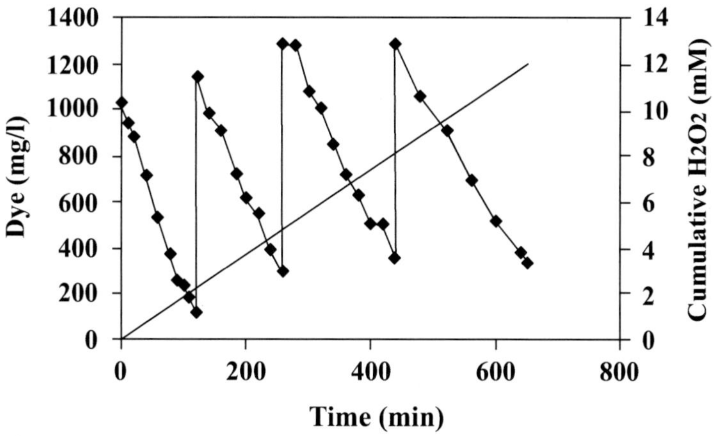

Fig. (54). Change in the dye concentration (♦) and cumulative H_2O_2 concentration shown as a straight diagonal line in the experimental condition C4 [62].

2.4. Discussion

One of the major concerns in DyP application is its inactivation by H_2O_2 because the high input or rapid addition of H_2O_2 inactivates several other enzymes [61, 118, 119, 196-199, 200, 2001]. Therefore, to reduce DyP inactivation, selection of reducing substrates, supply of low concentration of H_2O_2, or continuous supply of low level of H_2O_2 have to be considered.

In this chapter, the term, turnover capacity has been introduced to evaluate the efficiency of DyP under different H_2O_2 additions by measuring both the amount of decolorized dye by DyP and the decolorization rate of DyP. Four kinds of experimental H_2O_2 supply modes of C1-C4 have been designed [62].

Table **30** summarizes the results. When the turnover capacity is high and the decolorization rate is high, the high stability of DyP against H_2O_2 is reflected. Actually, the supply of a high concentration of the dye and the low ratio of H_2O_2/dye led to higher stability of rDyP against H_2O_2, reflecting an increase in turnover capacity of 5.76 in C2 to 14.3 in C3 even under similar H_2O_2 addition modes of C2 and C3. This result can also clearly be seen from the data of the increased decolorized amount of the dye of 2,875 mg/l dye in C3 from 1,152 mg/l in C2. The highest turnover capacity of 20.4 was obtained in C4 mainly because a low ratio of H_2O_2/dye was maintained by continuous fed-batch addition of H_2O_2. This concludes that the patterns of H_2O_2 addition and supplied dye concentration are key factors in protecting rDyP from inactivation by H_2O_2.

Table **29** shows that the A/D ratio, *i.e.*, the total amount of added H_2O_2 (A) to the total amount of added dye (D), changed from 1 in Cl to 1.9 in C4, and consequently, the amount of the decolorized dye increased. The ratio, A/E, of the total amount of added H_2O_2 (A) to the total amount of treated dye (E) exhibited a similar tendency to the ratio of A/D. This may be partly because, during the reaction time, by-products produced from dye-degradation by rDyP may have caused more consumption of H_2O_2, especially after the re-addition of H_2O_2. This is shown in Fig. (**51**).

Decolorization rates are also important for efficient enzyme reactions. When the average decolorization rates of dye, RBBR, were compared among four experimental conditions, the decolorization rate in C1 was the highest at 14.3 mg-dye/1/min. However, the inactivation of rDyP in C1 was also the largest, mainly because of the excessive supply of H_2O_2. Decolorization rates in C2 were higher than those in C3 and C4 mainly because at high initial ratios of H_2O_2/dye in C2, the formation of intermediate compound I was fast, resulting in a high decolorization rate at the expense of rDyP inactivation, as shown in Table **30**. Decolorization rates in the first fed-batch in C3 were similar to that in C4. However, the average decolorization rate of 2.57 in C4 was higher than 2.35 in C3, indicating that continuous fed-batch addition of H_2O_2 in C4 induced high turnover capacity by maintaining relatively high decolorization rates. As a result, the total amount of the dye treated was maximum in C4, as shown in Table **29**.

Some other fungal cultures decolorized RBBR. *Pleurotus ostreatus* decolorized 90% of 31 mg/1 of the dye [165] and *Irex lacteus* decolorized 95-100% of 150 mg/1 of the dye [202]. However, efficiencies of dye decolorization by these fungal cultures were relatively low because the high affinity of direct dye adsorption onto fungal cells induced longer cultivation time and lower efficiency [162, 203].

Table **31** shows RBBR decolorization by laccases from different fungi. The data from the rDyP application in the first batch operation in Cl to C4 are also presented. It is obvious that the decolorization ratio and the specific decolorization rate of the dye by rDyP were significantly higher than those by laccases [204, 205]. Moreover, the use of rDyP was simpler because most laccases require mediators to decolorize the dye [155, 206].

In the previous chapter, the possibility of the production of a high concentration of rDyP was demonstrated. By using rDyP of 5,000 U/1 with a turnover capacity of 20.4 in C4, one liter of crude rDyP was possible to decolorize 102g of the dye, RBBR. Therefore, a high concentration of non-purified crude rDyP can be a useful candidate for development as a practical catalyst in colored wastewater treatment.

Table 31. Comparison of RBBR decolorization by rDyP and laccases [62].

Enzyme (Species)	Enzyme Activity (U /ml), *(A)*	RBBR (mg/ml), *(B)*	Decolorization Ratio (%), *(C)*	Time (h), *(D)*	Specific Decolorization Rate (mg /h/ U), *(B×C)/(A×D)*	References
Laccase[a] *(T. versicolor)*	1.39	0.019	31	0.13	0.032	[187]
Laccase[a] *(Aspergillus)*	10	0.02	90	0.33	0.005	[148]
Laccase[a] *(L. edodes)*	5	0.0313	90	1.5	0.0037	[149]
Laccase[a] *(P. pinsitus)*	2	0.02	59	16	0.00036	[122]
Laccase[a] *(T. trogii)*	0.2	0.1	97	0.83	0.58	[192]
rDyP *(A. oryzae)*	0.2	0.15	95	0.17	4.2	This study, C1
	0.2	0.15	100	0.2	3.8	C2[b]
	0.2	0.930	84	1.66	2.35	C3[b]
	0.18	1.03	90	2	2.57	C4[b]

[a] One unit of laccase activity was defined as 1 μmol 2,2[1] -azino- bis(3- ethylbenzothiazoline- 6-sulfonic acid) (ABTS) in 1 min.

[b] Specific decolorization rates were valid only calculated for the fed-batch in C2, C3, and C4.

CONCLUSION

For DyP or rDyP application in practical processes, direct use or reuse of DyP and inactivation of DyP by H_2O_2 are the main problems that have to be solved. Immobilization of DyP is one of the potential ways to overcome these problems and maintain the high catalytic efficiency of rDyP. As immobilization of rDyP using conventional techniques was unsuccessful, chemically synthesized silica-based mesoporous materials, FSM-16 and AISBA-15, were introduced as a new carrier for immobilization. By repeated-batch dye decolorization using newly immobilized rDyP, overall efficiency defined as adsorption efficiency x activity efficiency for the materials was maximum at pH 5 for FSM-16 and at pH 4 for MISBA-15. Advantages of DyP immobilization on these mesoporous materials included the mild immobilization conditions and efficient physical adsorption of DyP. Thus, immobilized rDyP and massive production of rDyP provide the possibility for enzymatic treatment of colored wastewater. On the other hand, for optimization of the treatment system, minimization of both leaching from these materials and inactivation of the immobilized rDyP by H_2O_2 are essential.

With respect to the decolorization activity of rDyP, the inactivation of rDyP by an excessive supply of H_2O_2 has been found to be the most serious problem. In order to assess the high potential of rDyP, the turnover capacity was introduced. Under conditions of H_2O_2-mediated rDyP inactivation, four modes of addition of H_2O_2 were attempted, and turnover capacities were compared. As a result, continuous fed-batch of H_2O_2 and stepwise fed-batch of the dye provided the maximum turnover capacity of 20.4 by minimizing rDyP inactivation. At this turnover capacity, one liter of crude rDyP solution with 5,000 U activity could decolorize 102 g of the dye RBBR.

CHAPTER 6

Application of Dec 1 for Decolorization of Other Colored Substances

Abstract: Molasses is a primary carbon source, especially in the microbial industry; however, molasses includes many colored substances, like melanoidins, which become concentrated by the Maillard reaction after sterilization [41]. Thus, these remain in the Molasses Wastewater (MWW) after use. For effective treatment of MWW, biological methods are attracting attention. Section 1-10, in the previous Chapter 3, showed molasses to be available as a carbon source for the growth of Dec 1, and that partial color removal of molasses by Dec 1 was possible. The enhanced color removal of molasses by Dec 1 was conducted using a jar fermenter system consisting of fan-type agitators and a pressure swing adsorption oxygen generator. The oxygen-enriched air supply was effective not only in obtaining the highest decolorization degree of molasses, but also the highest activity of peroxidase, DyP for the decolorization of several dyes.

By simultaneous decolorization of molasses and an anthraquinone dye, RB5, the degree of decolorization of molasses reached 87%, and thus, the maximum decolorization rate of the dye, RB5, was achieved. However, the decolorizing activity of purufied DyP toward molasses was significantly lower than that of culture broth of Dec 1 due to the inhibitory effects of molasses on DyP, but the inhibition was reduced in the progress of degradation of molasses by growing Dec 1 concentration. Dec 1 degraded molasses containing substances with a wide range of molecular weights prepared by ultrafiltered fractions of molasses. As Dec 1 was not able to utilize sucrose, sucrose in the molasses was hydrolyzed with invertase to utilize all sugars in molasses. As a result, the decolorization of molasses and rate of decolorization of the dye, RB5, by Dec 1 reached the highest level.

A long and stable decolorization of molasses was attempted using both suspended and immobilized cells of Dec 1. In semi-batch cultivation using suspended cells of Dec 1, 80% decolorization of molasses and a stable DyP activity were maintained for approximately four weeks. When repeated batch cultivation of Dec 1 cells immobilized on polyurethane foam was applied, a longer and stable decolorization of molasses as well as stable DyP activity lasted for more than eight weeks.

Dec 1 was applied for the decolorization of kraft pulp bleaching effluent, abbreviated as E-effluent when glucose was supplemented. The color removal of E-effluent and the reduced amount of Absorbable Organic Halogens (AOX) reached 78% and 43%, respectively. The average molecular weight of colored substances in molasses was reduced to less than 3000 from the original 5600. The contribution of extracellular

Makoto Shoda
All rights reserved-© 2024 Bentham Science Publishers

enzymes, such as Peroxidase (DyP) and Manganese Peroxidase (MnP), to the decolorization of the kraft pulp bleaching effluent was observed in the later stage of decolorization.

Dec 1 decolorized up to 72% of Oxygen-delignified Bleaching Effluent (OBE). Biobleaching of Oxygen-delignified Kraft Pulp (OKP) was conducted at 2% pulp concentration. The brightness of OKP increased by 13% and the kappa value of OKP decreased by 4 points only for 3 days. However, at 25% of pulp concentration, the brightness of OKP increased only by 4% and the kappa value decreased by 3 points during a 12-day incubation period mainly because of oxygen limitation. When the culture after OBE decolorization was used for bleaching of OKP, the brightness of OKP increased to 62.7% at 2% pulp concentration. In the decolorization and biobleaching, the involvement of DyP and MnP was confirmed. From these results, the potentiality of Dec 1 for the decolorization of kraft pulp wastewater and biobleaching of kraft pulp in paper mills can be observed.

Keywords: Absorbable organic halogens, Immobilization on polyurethane foam, Kraft pulp bleaching effluent, Kappa value, Molasses, Oxygen-delignified bleaching effluent, Oxygen-delignified kraft pulp, Pressure swing adsorption oxygen generator, Repeated-batch cultivation, Ultrafiltered fractions of molasses.

1. DECOLORIZATION OF MOLASSES BY DEC 1

1.1. Methods

1.1.1. Decolorization Methods

The crude molasses used contained 33% sucrose, 6.5% glucose, 7.5% fructose, and 0.57 g Kjeldahl nitrogen/kg (w/w) [42]. The crude molasses was diluted to 50 g/1 and 20 g of glucose, 0.5 g of ammonium tartrate, 1 g of KH_2PO_4, and 0.5g of $MgSO_4 .7H_2O$ (per 1 liter) were supplemented. This medium was referred to as the molasses medium. The dye used for measuring the DyP activity was Reactive Blue 5 (RB5).

The decolorization degree of the molasses (%) was calculated by the methods reported previously [44, 46, 121].

A conventional jar fermenter (nominal volume=7 liter) was modified by removing some gears in the fermenter to simplify the inside and to avoid adhesion of Dec 1 to the solid parts. Three fan-type propellers were prepared to agitate the culture broth homogeneously and two top propellers and one bottom propeller were fixed for the broth to flow in opposite directions.

The sampling of the culture broth was aseptically carried out using a pump and the loss of the broth was supplemented by a fresh molasses medium, which was fed by another pump. Oxygen-enriched air prepared by a pressure swing adsorption oxygen generator avoided oxygen limitation. By a Dissolved Oxygen (DO) sensor, DO concentration was maintained.

1.1.2. Determination of the Activity of DyP

The decolorizing activity of DyP in the supernatant prepared from the culture broth of Dec 1 was determined in a mixture of the supernatant, H_2O_2, and dye, RB5, in citrate buffer (pH3.2) at the maximum absorbance wavelength of 600 nm of the dye [1]. The decolorizing rate was expressed as a concentration difference for the dye between two different times in the culture broth per unit time, and expressed as mg /1/ min.

1.2. Results

The effect of aeration and agitation in a reactor was investigated. The air supply rate was fixed at 0.4 vvm (volume of air/volume of medium/minute), and the agitation speed was changed from 150 to 250 rpm (rotation per minute). At 150 rpm, the growth of Dec 1 was slow, which may be mainly due to the low oxygen transfer rate. When the agitation speed increased to 250 rpm to supply enough oxygen, large pellets of Dec 1 with 2 mm in diameter were formed, and the decolorization rate of the molasses reduced significantly. At 180 rpm, the small mycelial pellets of Dec 1 were homogeneously dispersed and the enhancement of the decolorizing activity and DyP activity was observed. At a fixed agitation speed of 180 rpm, different air supplies were tested to obtain efficient decolorization conditions. Fig. (**55**) shows the results. The decolorization degree of molasses (Fig. **55a**) and DyP activity (Fig. **55b**) were the highest when enriched 60% oxygen was supplied at 0.4 vvm. Under this condition, the absorbance of the cell-free culture broth at 475 nm changed from 2.5 at 0 day to 0.5 at 10 days, suggesting the steady decolorization of molasses.

The growth of Dec 1 (Fig. **55c**) at 60% oxygen supply or air supply at 0.4 vvm was similar. The apparent growth in the later days reflected the fast consumption of glucose, as shown in Fig. (**55d**). Although the activity of DyP increased in 6 days, the contribution of DyP to the decolorization of molasses was found to be small. The details are provided in the next section 6-2.

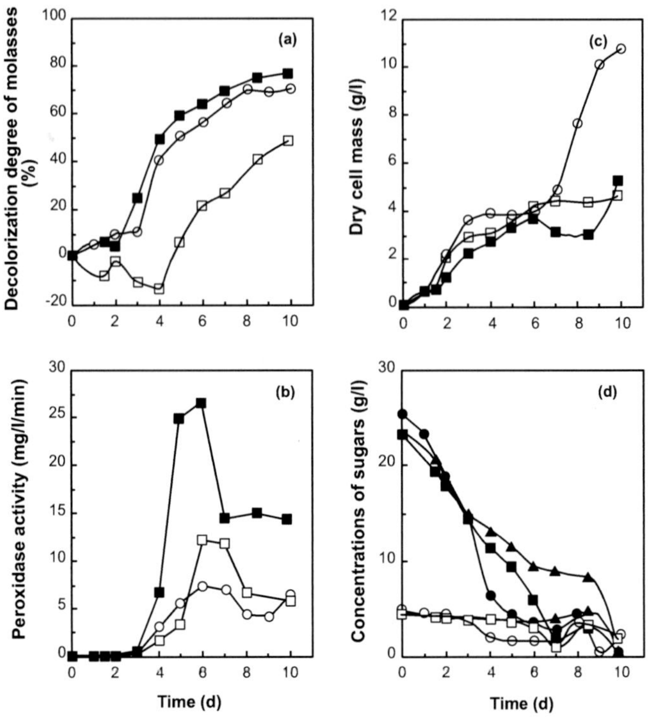

Fig. (55). Molasses decolorization by Dec 1. Decolorization degree of molasses (**a**); peroxidase, DyP activity (**b**); growth as dry cell mass (**c**); change in glucose and fructose concentrations (**d**). Symbols of (a), (b) and (c): □:0.4vvm of air, ○: 0.6vvm of air, ■: 0.4 vvm of 60% O_2-fortified air; symbol of (d): glucose (▲: 0.4vvm of air, ●: 0.6vvm of air, ■: 0.4vvm of 60% O_2-fortified air), fructose (Δ: 0.4vvm of air, ○: 0.6vvm of air, □: 0.4vvm of 60% O_2-fortified air) [46].

When nitrogen concentration increased, the decolorizing ability of Dec 1 drastically decreased mainly because the carbon source and nitrogen accelerated the growth of Dec 1. However,this did not lead to DyP production , reflecting that DyP production is non-growth-associated.

Compared to a fungus, *P. chrysosporium*, which was used for decolorization, Dec 1 exhibited a high growth rate, with lower sensitivity to shear stress and tolerance to high dye concentrations [1]. The possibility of the simultaneous decolorization of molasses as a carbon source and a dye under oxygen-enriched conditions is presented in the following section.

One disadvantage of Dec 1 is that this fungus cannot consume sucrose. For the full utilization of molasses using Dec 1, three methods were employed, including the hydrolysis of sucrose by chemical or biological reactions, the genetic transformation of Dec 1 with a gene encoding invertase, or a mixed culture with sucrose-hydrolyzing microorganisms. The result of invertase hydrolyzation is presented in the next section.

2. SIMULTANEOUS DECOLORIZATION OF MOLASSES AND A DYE

2.1. Methods

The strain used, molasses and dye used, medium composition, and decolorization assay have been described in a previous section.

2.1.1. Inhibitory Effect of a Fraction of the Molasses Medium on the Activity of Purified Peroxidase, DyP

At each sampling time, the cell-free culture broth was ultrafiltered and the filtrate fraction with a molecular weight of less than 10,000 was concentrated by freeze-drying. The activity of DyP was determined by the method described in the previous section. The degree of inhibition has been defined as the relative DyP activity value measured in the presence of the fraction relative to that in the absence of the fraction.

2.1.2. Decolorization of Ultrafiltered Fractions of Molasses by Dec 1

The diluted molasses concentrated by ultrafiltration to obtain fractions with a molecular weight of more than 10,000 and with a molecular weight of less than 10,000 were freeze-dried. Each of the two sterilized fractions was mixed with a culture of Dec 1 and the decolorization of each fraction was monitored at 475 nm.

2.1.3. Decolorization of Molasses and the Dye in Invertase-treated Molasses Medium

Molasses was mixed with invertase and hydrolyzed. Complete inversion of sucrose to glucose and fructose was confirmed by HPLC. This solution was mixed with the dye and mineral solution. Dec 1 was cultivated in this solution, and decolorization of molasses and the dye was determined.

2.2. Results

2.2.1. Growth of Dec 1 in Molasses

Fig. (**56**) shows changes in the growth of Dec 1, pH, and sugar concentrations in 40 g/l of molasses. The pH value was stably maintained around 4-5. Glucose was utilized first, and then fructose was consumed, but sucrose remained unconsumed, indicating Dec 1 has no ability to utilize sucrose. Actually, the degree of decolorization of the molasses was not significant, mainly because the available sugar concentrations in the molasses were relatively low. Therefore, to enhance the decolorization activity of Dec 1, glucose, ammonium tartrate, KH_2PO_4, and Mg $SO_4 \cdot 7H_2O$ were supplemented. The addition of glucose was determined to make the same concentrations of sugars available to Dec 1 in the molasses when the molasses was hydrolyzed by invertase.

2.2.2. Decolorization of Molasses and a Dye in a Molasses Medium

Fig. (**57**) shows the decolorization of molasses and a dye in a molasses medium containing additional glucose and inorganic nutrients. Fig. (**57a**) shows the change in decolorization degree of molasses and decolorization rate of the dye, RB5, in the culture broth, and DyP activity in the cell-free culture broth. Fig. (**57b**) shows the change in glucose and fructose concentrations. The changes in pH, demonstrated in Figs. (**56** and **57**), were similar. The color change of molasses started after 3 days of incubation and gradually increased to 87% at 12 days when glucose and fructose were exhausted.

The maximum decolorization rate of RB5 by the culture broth of Dec 1 was 228 ppm/h in 7 days, as shown in Fig. (**57a**). This value was about twice as large as that observed previously in the PD medium [1]. The decolorization activity of crude DyP solution for the dye appeared after 4 days and reached a maximum value of 75 ppm/min in 7 days. The value was about 20 times larger than that achieved with the cell-free culture broth of 3.8 ppm/min.

Further purification of DyP was conducted using the 7-day culture broth, as shown in Fig. (**57a**), and the obtained DyP was used for decolorization of the dye. The activity of purified DyP increased to 668 ppm/min, which was about 200 times higher than the apparent cell-free culture broth activity of 3.8 ppm/min. This indicates that the dye decolorization in the molasses medium was adversely affected by molasses inhibition.

After the molasses was ultrafiltered into the two fractions with a molecular weight of over 10,000 and under 10,000, Dec 1 was cultivated in a PD medium containing each fraction. Fig. (**58**) shows that the degree of decolorization was 87% and 78% for the two fractions with a molecular weight fraction of over 10,000 and a molecular weight fraction of under 10,00, respectively. Crude molasses contained colored substances having a broad range of molecular weight, but after decolorization over the 12-day period, a significant decrease in absorbance was observed, showing Dec 1 to possess the ability to decolorize a wide range of fractions in the molasses.

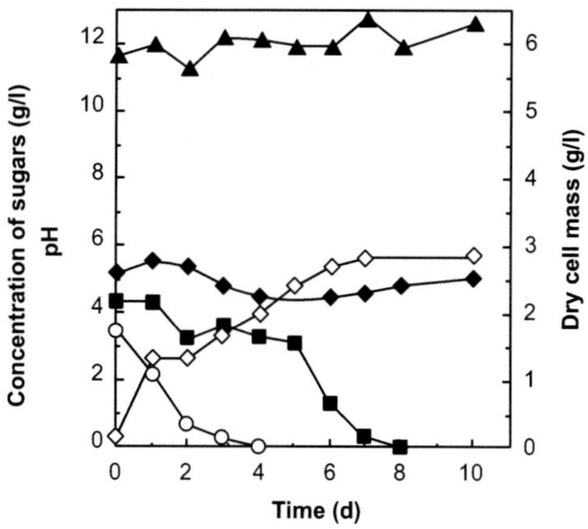

Fig. (56). Changes in growth (◇), pH (♦) and sugar concentrations (glucose (○), fructose (■) and sucrose (▲) when Dec 1 was grown in molasses medium [21].

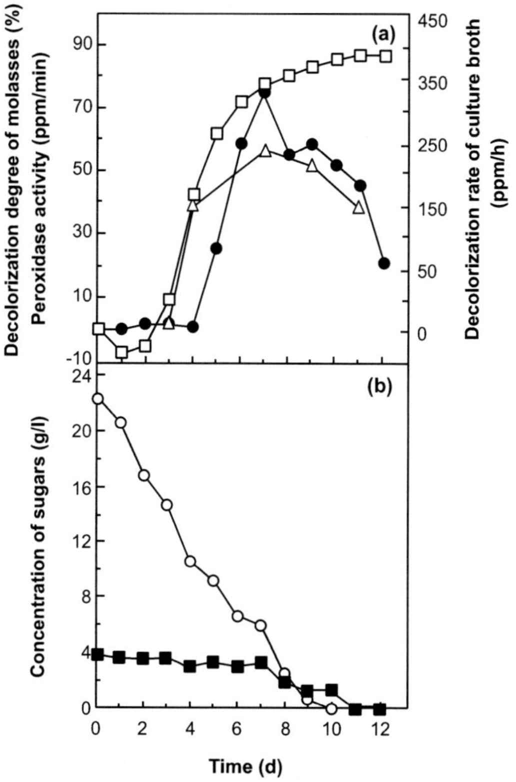

Fig. (57). Decolorization of molasses and the dye, RB5, by Dec 1 grown in the molasses medium. (**a**) Decolorization degree of molasses (□), decolorization rate of the culture broth for the dye, RB5 (Δ), and DyP activity in the cell-free culture (•); (**b**) Glucose (○) and fructose (■)concentrations [21].

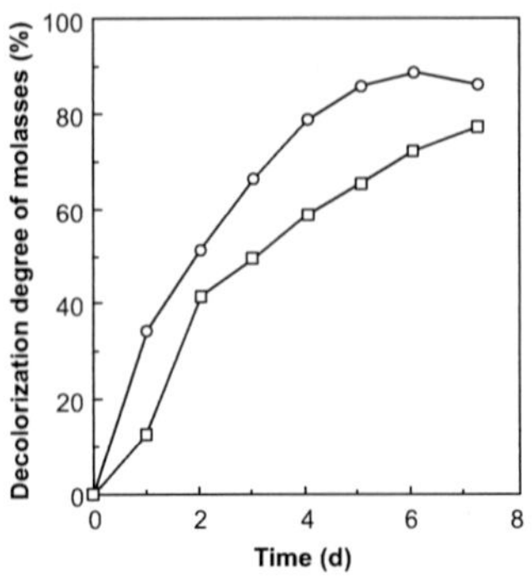

Fig. (58). Decolorization of fractionated molasses by Dec 1 [21]. (○) The fraction including substances with a molecular weight of over 10,000. (□) Fraction including substances with a molecular weight of less than 10,000.

2.2.3. Change in DyP Activity During the Growth of Dec 1 in Molasses Medium

The growth of Dec 1 in the molasses medium tended to show a longer lag time compared to that in the PD medium, and the expression of DyP activity started a few days later. As the molasses contains some inhibitory substances, the change in DyP activity in the molasses was investigated in detail. The culture of Dec 1 cultivated in the molasses medium was centrifuged periodically to obtain supernatant and a fraction containing molecular weight of less than 10,000 was separated. Fig. (**59**) shows the relative inhibition degree of DyP activity measured using the fraction that contained a molecular weight of less than 10,000. Until 5 days of cultivation, the inhibition of DyP activity was significant, but the inhibition was reduced after 6 days when the growth of Dec 1 became active. This proves that Dec 1 degraded inhibitory substances during the growth and inhibition of DyP activity became less significant in the later period of cultivation.

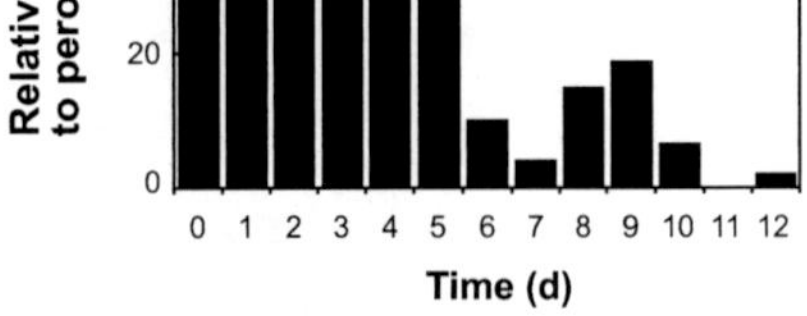

Fig. (59). Relative degree of inhibitory effects of fractionated molasses medium on the activity of DyP at each cultivation time [21].

2.2.4. Decolorization of Molasses by Purified DyP

In order to clarify the adverse effect of substances derived from molasses on the activity of DyP, the ability of purified DyP to decolorize molasses was assessed at optimal pH 3.2. At pH 3.2, the decolorization of RB5 by DyP was previously found to be optimal [1] and almost no significant decrease in decolorizing activity was observed. However, at pH 5, the ability decreased only to 12%, and at pH 7, a much slower decolorization than at pH 5 occurred, suggesting the contribution of DyP to the decolorization of molasses to be low.

2.2.5. Decolorization of Molasses and the Dye in the Invertase Hydrolyzed Medium

As Dec 1 cannot degrade sucrose, sucrose in the molasses remained undegraded during decolorization. When sucrose in the molasses was hydrolyzed by the enzyme invertase, about 12 g/l each of glucose and fructose was produced, and these two sugars were utilized by Dec 1. Fig. (**60**) shows the details of the decolorization of Dec 1 in a molasses medium that was hydrolyzed by invertase. During the 12-day cultivation, the degree of decolorization of molasses reached 87% by consuming the two sugars. The value was almost similar to that shown in Fig. (**57**).

2.3. Discussion

So far, no microbial study has been reported to decolorize both the intact molasses and dyes. A newly isolated fungus, Dec 1 utilized sugars as carbon sources in the crude molasses for the growth and decolorized the molasses significantly, as shown in Fig. (**57**). Although the color of molasses intensified due to melanoidins after sterilization, decolorization of the molasses by Dec 1 proceeded steadily [21].

When the anthraquinone dye, RB5, was added to the molasses medium, a maximum decolorization rate of 280 ppm/h was obtained in 7 days. This value indicates that about 10 g/l of the dye decolorized in one day. As highlighted in Chapter 2, Dec 1 decolorized the dye at an initial concentration of about 11 g/l without any decline in decolorizing rate, indicating that Dec 1 exhibited tolerance to high concentrations of dyes that are generally inhibitory to microbial activity [1]. The dye decolorization ability of Dec 1 was noted even in the mixture of molasses.

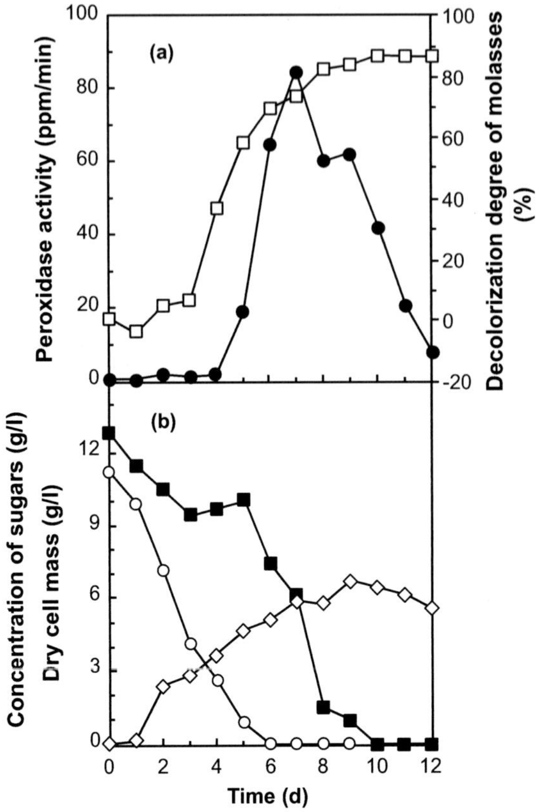

Fig. (60). Decolorization of molasses by Dec 1 that was hydrolyzed by invertase. (**a**) Decolorization of molasses (□) and DyP activity of the cell-free culture (●); (**b**) Glucose (○), fructose (■), and growth of Dec 1 (◇) [21].

Dec 1 produced DyP, one of the enzymes involved in decolorization of dyes [1]. Despite the high dye decolorization activity of crude DyP, the color change of the dye by the Dec 1 culture broth was not noticeable. This was partly attributed to the presence of inhibitory substances in the molasses because the activity of purified DyP was significantly inhibited by the culture of Dec 1, especially during the initial stages of growth, as shown in Fig. (**59**). Although the inhibitory substances had not been identified, the fraction containing substances with a molecular weight of less than 10,000 exhibited to be responsible for the inhibition. Although some enzymes involved in the decolorization or mineralization of melanoidins were reported [198, 199], they exhibited only 25% decolorization activity for natural or synthetic melanoidins under optimum conditions [207 - 209]. In this experiment, initially DyP showed only 10% decolorization of molasses, but later decolorization level increased.

The dye decolorization activity declined significantly after 7 days, as shown in Fig. (**57**). The reason for this was not clear. When the glucose concentration decreased to less than 2 g/l after 8 days of cultivation, significant deterioration of the dye decolorization activity was observed, but when fresh glucose was added, the recovery of DyP activity was confirmed with the growth of the cells, as described in Chapter 2. Although the adverse effect of molasses on DyP activity was clear, a continuous supply of sugars showed continuous production of the DyP to be possible by maintaining dye decolorization activity. From the data, it can be inferred that cultivation by fed-batch culture or continuous culture can be appropriate for efficient utilization of Dec 1 for decolorization.

The hydrolysis of sucrose in the molasses by invertase treatment produced a mixture of glucose and fructose, and Dec 1 utilized them effectively for dye decolorization. This indicates the availability of sucrose in molasses to be a key point for the efficient activity of Dec 1. Therefore, the following two experiments can be conducted to increase the applicability of Dec 1 to decolorization: (1) genetic transformation of Dec 1 with a gene encoding invertase, or (2) a mixed culture of Dec 1 with invertase producing microorganisms.

When the hydrolyzed medium containing ammonium concentration at 0.5 g/l as a nitrogen source was used for decolorization of the dye, RB5, the decolorization degree was significantly high compared to that with those containing 2 or 5 g/l ammonium. Actually, at 2 or 5 g/l of ammonium tartrate, all sugars were exhausted within 3-4 days, but no decolorization activity occurred. This indicates that carbon and nitrogen sources rather work for the growth of the cells than the production of enzymes, and thus, a medium containing a lower nitrogen concentration should be preferable for effective decolorization by inducing the production of second metabolites. These types of results have also been reported for other lignin-degrading fungi [5, 94]. Moreover, the higher nitrogen concentration caused the adsorption of the dye on the surface of the fungi, and this delayed the decolorizing rate, indicating that the cell wall may be changed according to nitrogen concentration.

As a dye-decolorizing fungus, *P. chrysosporium* [5, 98, 99, 210] is one of the extensively studied fungi. Dec 1 is different from *P. chrysosporium* in that the growth rate of Dec 1 is higher, Dec 1 is less sensitive to shear stress and higher dye concentrations, and Dec 1 occurs widely in soil and water. For more efficient utilization of Dec 1, the immobilization of Dec 1 cells can provide an advanced method.

3. SIMULTANEOUS DECOLORIZATION OF MOLASSES AND A DYE BY SUSPENDED AND IMMOBILIZED DEC 1 CELLS

3.1. Methods

The strain, molasses, and PD medium used in this experiment have been described previously. The structure of a conventional jar fermenter (working volume 5 liter) used for repeated-batch cultivation was described in a previous section, 6-1. Batch cultivation was initially conducted until 80% molasses was decolorized, and then, the culture broth was sedimented to leave the fungal cells at the bottom of the reactor. The liquid was removed from the reactor and the same volume of fresh molasses medium was added, and the cultivation proceeded until decolorization of the molasses reached 80%. This procedure was repeated for 41 days. A 60% oxygen-enriched air produced by a pressure swing adsorption oxygen generator was supplied to secure enough Dissolved Oxygen (DO) concentration in the reactor. DO concentration and pH were monitored continuously.

3.1.1. Decolorization Rate of the Dye, RB5

Evaluation of the direct decolorization degree of RB5 was difficult because the black-colored molasses interfered with the reaction. DyP activity was instead used to assess the decolorization ability of the dye, RB5, by using the culture broth. The culture broth sample was mixed with RB5 [1, 46] and H_2O_2, and decolorization of the dye was determined by measuring the absorbance of RB5. The decolorization rate of RB5 was expressed as ppm of the dye/min.

3.1.2. Immobilization of Dec 1

Polyurethane foams were selected as a carrier of immobilization because of their lightness and porosity [211, 212]. The properties of the polyurethane foam are as follows: bulk density, 35 kg/ m^3; indentation load deflection, 12.5 kg; tensile strength, 0.7kg/cm^2; elongation, 150%; and compression set, 5%. The cells of Dec1 immobilized on the polyurethane foam carrier are shown in Fig. (**61**).

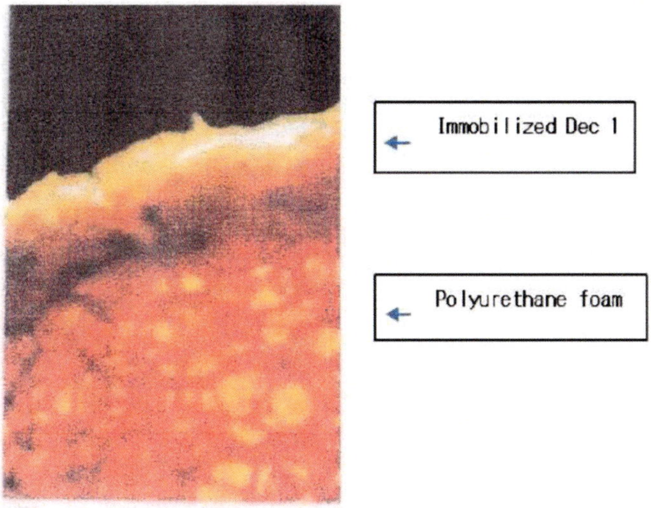

Fig. (61). Dec 1 cells immobilized on the polyurethane foam [47].

3.2. Results

3.2.1. Decolorization of Molasses by Suspended Dec 1

The change in the decolorization degree of the molasses and decolorization rate of RB5 in repeated-batch cultivation of Dec 1 is shown in Fig. (**62**). After an initial batch culture for 10 days, repeated-batch operation was started. The decolorization degree of molasses reached 80% and it significantly declined later. The DyP activity expressed as the decolorization rate of RB5 reached a maximum after 18 days and started to decline. During the experimental period, enough glucose concentration was detected and a sufficient DO concentration was confirmed. Therefore, the reason for the deterioration of the decolorization activity was clarified as follows: the culture broth at 41 days was centrifuged and the precipitated cells of Dec 1 were spread on PDA medium and cultivated to obtain a spore suspension of the cells. The spore suspension was then inoculated into a fresh molasses medium and cultivated. After 8-day cultivation, the degree of molasses decolorization and DyP activity were measured. The collected cells showed only 25% of the molasses decolorization activity of the original cells. When the original Dec 1 cells were cultivated in the cell-free culture broth taken at 41 days, the degree of molasses decolorization and DyP activity were also significantly low. This indicates that molasses components after the long cultivation time inhibited the cell activities of Dec 1 or some enzymes associated with molasses degradation might have become unstable [198, 199]; however, no specific enzymes were clarified.

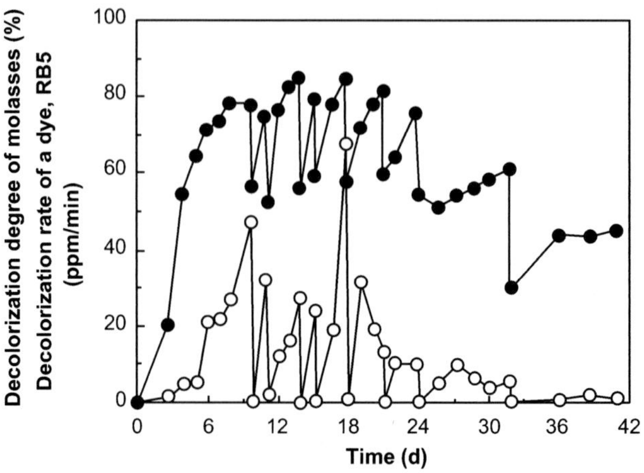

Fig. (62). Decolorization by the repeated-batch culture broth of Dec 1. Decolorization degree of molasses (●) and decolorization rate of RB5 (○) [47].

3.2.2. Decolorization of Molasses by Immobilized Dec 1

Repeated-batch cultivation with Dec 1 immobilized on a polyurethane foam carrier was conducted. Fig. (**63**) shows the results. 80% of the decolorization degree of molasses and about 15 ppm/min of decolorization rate were stably maintained for 61 days, as shown in Fig. (**63a** and **63b**). The pH fluctuated approximately every 2 days, reflecting the fresh medium addition after the decolorization degree of molasses reached 80%. The estimated cell mass of Dec 1 on the polyurethane carrier was 3 g-cell/g-polyurethane. This value was 3-fold higher than the data reported previously [213]. As the suspended cell mass was significantly small compared to immobilized cells, the effectiveness of immobilization of the cells was obvious.

During the suspended Dec 1 cultivation, homogeneous dispersion of the cells was possible by controlling agitation speeds. However, the enhanced agitation produced larger pellets of Dec 1, leading to an oxygen-limited environment. The supply of oxygen-enriched air was effective in maintaining high DO, but the rigid control of the pelleted size of Dec 1 by agitation or aeration was difficult. Thus, Dec 1 immobilized on solid support was introduced for controlling the activity of the cells.

The advantages of immobilization of the fungus included short retention time, easy recovery of the cells, and increased decolorization activity. Furthermore, the urethane foam matrix appropriately restricted the pelleted size of Dec 1 by employing the physical properties of the foam. To take advantage of immobilized cells, the need for careful operation has been reported to avoid a decrease in the maximum activity level [214, 215].

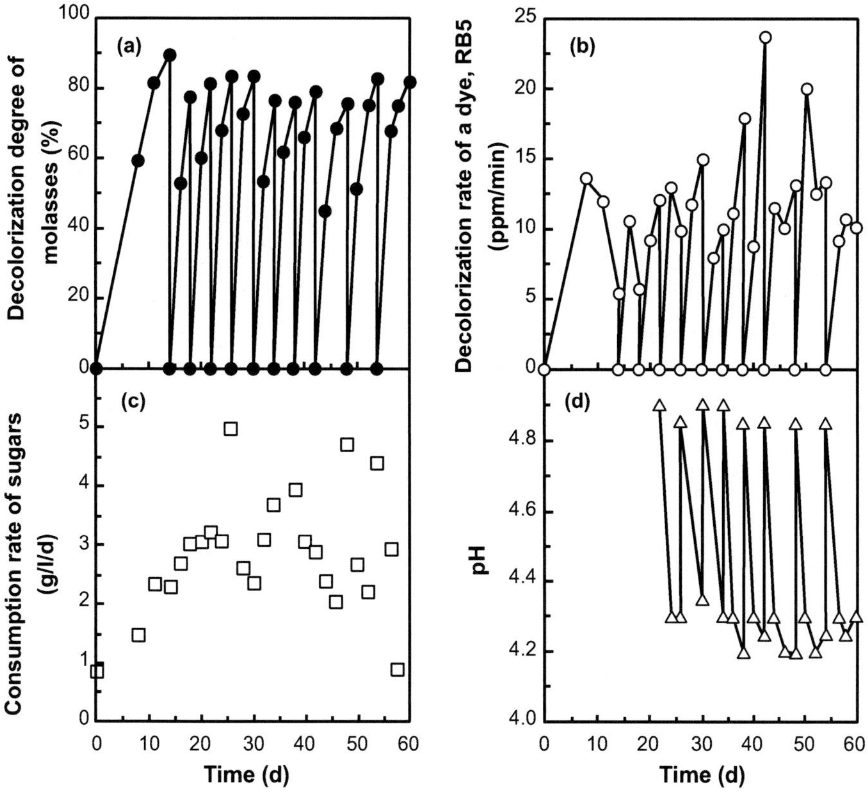

Fig. (63). Decolorization by Dec 1 cells immobilized on polyurethane foam in repeated-batch cultivation. Decolorization degree of molasses (●) (**a**), decolorization rate of RB5 (○) (**b**), the consumption rate of sugars of glucose and fructose (□) (**c**), and pH (Δ) (**d**) [47].

The selection of carbon sources is an important factor in the practical use of Dec 1. Although molasses is one of the most popular carbon sources, the issues of color removal and the inhibitory effects of molasses should be solved. The consumption of molasses as a carbon source and its decolorization were efficiently carried out by Dec 1, and also stable DyP activity was mediated by the immobilized Dec 1 cells. The results obtained here can be applied to different systems for dye decolorization, such as one-stage semi-batch systems or two-stage continuous systems.

Conclusively, by immobilizing Dec 1 cells, the adverse effect of molasses on the cell activity was avoided and a stable decolorization activity of Dec 1 was realized.

4. DECOLORIZATION OF KRAFT PULP BLEACHING EFFLUENT BY DEC 1

4.1. Methods

4.1.1. Strain and Preparation

The details of Dec 1, potato dextrose, PD medium, and GY medium have been described in previous sections [216].

4.1.2. Lignin Decolorizing Assay

Dec 1 spores were inoculated on a GY agar plate containing commercial lignin 1 g/1 and incubated for 2 weeks at 30°C.

4.1.3. Effect of pH on the Kraft Pulp Bleaching Effluent, E-effluent

The effluent from hardwood pulp treated with alkaline extraction was named kraft pulp bleaching effluent and abbreviated as E-effluent. The supernatant obtained from centrifugation of E-effluent was mixed with citrate buffer at different pHs (3.0, 4.5, and 5.0), and the absorbance of the supernatant at 465 nm against distilled water was measured to check for any differences.

4.1.4. Cultivation of Dec 1

Glucose, ammonium tartrate, KH_2PO_4, $MgSO_4 \cdot 7H_2O$, yeast extract, and Dec 1 spores were mixed and cultivated at 30°C.

4.1.5. Color Removal Method

The color removal of the E-effluent was determined by the standard method described by the Canadian Pulp and Paper Association. The values of color absorbance were changed into Color Units (CU) by using the following formula:

$$CU = 500 \text{ X } A_2/A_1$$

Where, A_1 is the A_{465} of a 500-CU platinum-cobalt standard solution (0.132) and A_2 is the absorbance of the E-effluent sample. The color removal (%) was defined as the ratio of CU of the culture sample to that of the initial medium. The average initial color unit of the E-effluent was determined beforehand as 4200.

4.1.6. Color Adsorption to Dec 1 Cells

Adsorption of colored substances in E-effluent to the Dec 1 cells often occurs before dye degradation by the cells of Dec 1. In order to evaluate accurate color removal, the value of color adsorption has to be determined. Colored substances adsorbed to the Dec 1 cells were removed by alkaline treatment when the maximum color removal time came. The CU of the removed solution was evaluated by the method described above as the mass of colored substances adsorbed to the cells [217].

4.1.7. Absorbable Organic Halogens (AOX)

Absorbable Organic Halogens (AOX) of E-effluent were determined using the AOX analyzer.

4.1.8. Gel Filtration of E-effluent

The distribution of the molecular weight of the colored substances in the E-effluent was determined with gel filtration column chromatography using 280 nm absorbance. The fraction range in filtration was obtained between 3000 and 80000 as globular proteins. The standard molecular size markers used are shown in Fig. (**68**).

4.1.9. Enzyme Assay

The activities of the following four enzymes were determined.

4.1.9.1. DyP

The assay of DyP activity has been described in detail in a previous chapter [20]. One unit of DyP activity was defined as the amount of the DyP that decolorized 1 μmol of the dye, RB5, at 30°C for 1 min.

4.1.9.2. Manganese Peroxidase (MnP)

Guaiacol was used as the substrate for MnP assay.

The detail of the method has been described in a previous chapter [216]. One unit of MnP activity was defined as the amount of MnP that reacted with 1 μmol of guaiacol at 30°C for 1 min.

4.1.9.3. Lignin Peroxidase (LiP)

Veratryl Alcohol (VA) was used as the substrate for the LiP assay [216. 223]. In the culture supernatant of Dec 1, VA and H_2O_2 were mixed, and the increase in absorbance at 310 nm was determined. The enzyme activity was calculated from the increase in absorbance at 310 nm. One unit of LiP activity was defined as the amount of LiP that reacted with 1 μmol of VA at 30°C for 1min.

4.1.9.4. Laccase (Lac)

2, 2'-Azino-di-(3-ethylbenzthiazoline)-6' sulfonate (ABTS) was used as the substrate for Lac by referring to a previous report [212, 218]. One unit of Lac activity was defined as the amount of Lac that reacted with 1 μmol of ABTS at 30°C for 1 min.

4.2. Results

4.2.1. Decolorization of Lignin by Dec 1

The degradation of lignin by Dec 1 was confirmed by the color change in the plate, as shown in Fig. (**64**).

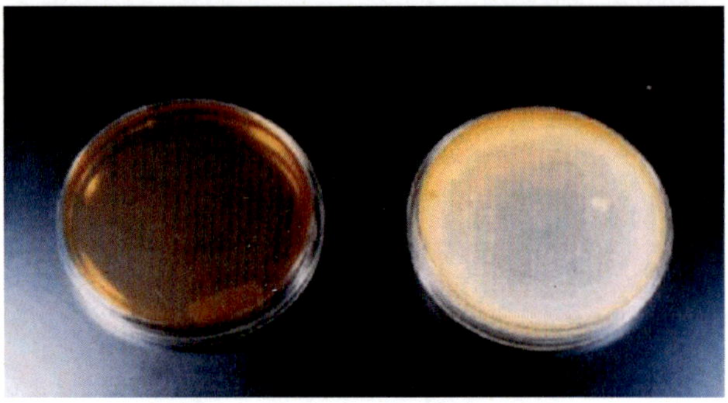

0 day 9 days

Fig. (64). Degradation and decolorization of 1000ppm lignin by Dec 1 in GY agar plates. Left: 0 day; right: after 9 days (prepared by the author).

4.2.2. Effect of pH on E-effluent

When the pH of the E-effluent was adjusted to 3.2, 4.5, or 5.0, the absorbances at 465 nm were 0.474, 0.494, and 0.512, respectively. As a small change in the absorbance at acidic pH was obtained, the pH difference exhibited no strong effect on the decolorization of the E-effluent.

4.2.3. Decolorization of the E-effluent

Fig. (**65**) shows the change in color removal and the pH of the E-effluent under different initial glucose concentrations. The maximum color removal was 78% at glucose concentration of 20 and 30 g/l. Neither color removal nor growth of Dec 1 was observed without glucose.

At glucose 5 g/1, color removal was 47%, but it decreased to 0 after 4 days, and pH increased, indicating exhaustion of glucose. The increase in pH was presumably because acidic by-products derived from the degradation of glucose were consumed by Dec 1. When Dec 1 was not inoculated into the E-effluent mixture, color removal was 15% at most, as shown in Table **32**.

Fig. (**66**) shows the change in the different initial glucose concentrations. It is obvious that glucose concentrations were closely related to the duration of the color removal, and color removal disappeared after glucose was exhausted in all cases.

Table 32. Color removal of E-effluent mixture without Dec 1 under different pHs [216].

Day	Color Removal (%)		
	At pH 3.0	At pH 4.0	At pH 5.0
0	0	0	0
3	10	4	1.0
7	15	4.5	1.5
10	14	4.3	1.6

Table 33. Relation between color removal and decomposed colored substances at various glucose concentrations [216].

Glucose Concentration of Medium (g/l)	(A) Initial CU	(B) CU of Culture Supernatant [a]	Color Removal (%) (A - B) /A × 100	(D) CU after Alkaline Treatment [b]	Ratio of Decomposed Colored Substances [c] (%) (A - B - D) /(A - B) × 100
5	4110	2310	44	1960	0
10	4200	1230	71	1140	62
30	4410	1140	74	320	90

CU. Color unit;
a: Culture times of 5, 10, and 30 g of glucose medium per liter correspond to 3, 4, and 9 days, respectively;
b: This value represents the amount of adsorbed colored substances to Dec 1, but not decomposed colored substances by Dec 1;
c: Minus value has been treated as 0.

4.2.4. Color Adsorption to Dec 1 Cells

Table **33** summarizes the relation between apparent color removal and net decomposed color removal at various glucose concentrations At a glucose concentration of 5g/l, 44% of the color removal was mainly due to the adsorption of colored substances to the Dec 1 cells because the ratio of decomposed color removal was 0%. In contrast, at 10 or 30g/l concentration, the ratio of net decomposed color removal increased to 62% and 90%, respectively.

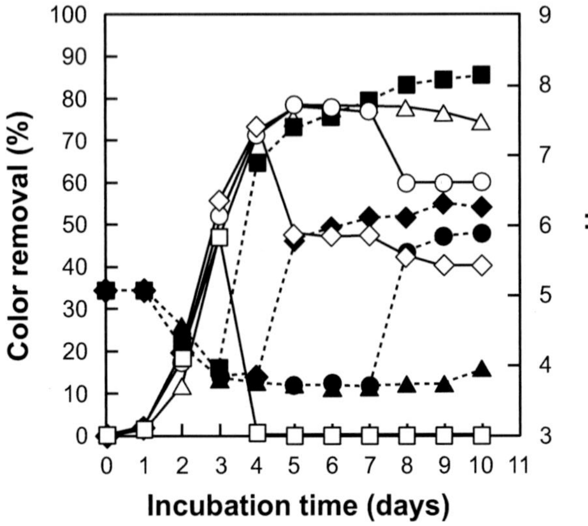

Fig. (65). Effect of different initial glucose concentrations on color removal and pH of E-effluent. Open and closed symbols represent color removal and pH, respectively [216]. (□ ■): 5g glucose/1; (◇◆)10g/1; (○●): 20g/1; (Δ ▲): 30g/l.

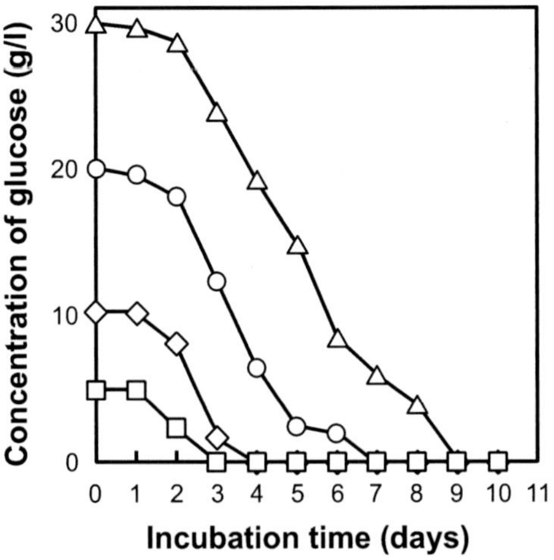

Fig. (66). Change in different initial glucose concentrations in E-effluent [216]. (□): 5g/1; (◇): 10 g/1; (○): 20 g/1; (Δ): 30 g/1.

4.2.5. Reduction of AOX

Fig. (**67**) shows the change in color removal and reduction of AOX of E-effluent. The reduction of AOX and color removal achieved 43% and 78% after 7 days, respectively. This value indicates that Dec 1 degraded a significant amount of aromatic halogens that were included in the pulp bleaching process.

4.2.6. Molecular Weight Distribution of the E-effluent Determined by Gel Filtration

Fig. (**68**). shows the molecular weight (MW) distribution of the colored substances of the E-effluent by gel filtration column chromatography. The MW of the main colored substance was estimated to be 5600 at the elution peak. After 7 days, MW of the colored substances decreased to less than 3000 because its elution point was the same as that of vitamin B.

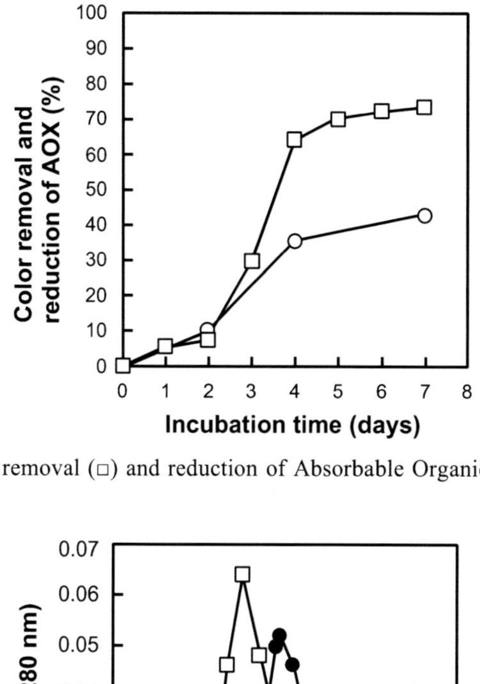

Fig. (67). Change in color removal (□) and reduction of Absorbable Organic Halogens (AOX) (○) of E-effluent [216].

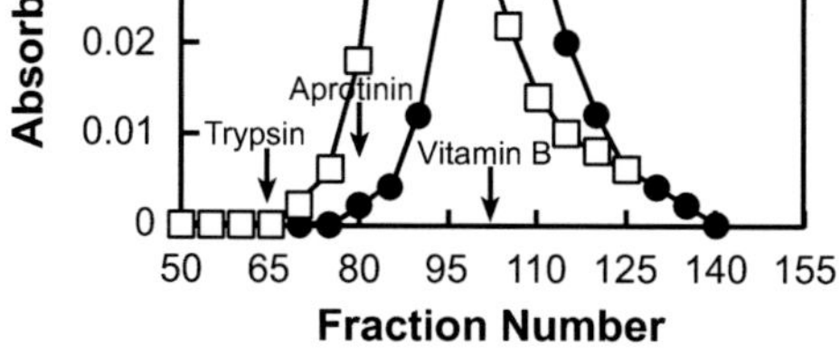

Fig. (68). Molecular weight distribution of E-effluent (□) and E-effluent treated with Dec 1 for 7 days (•). The elution points of trypsin (MW 24000), aprotinin (MW 6512), and vitamin B (MW1355) are indicated by arrows [216].

4.2.7. Enzyme Activity and Color Removal

Fig. (**69**) shows changes in color removal and enzyme activities when E-effluent was supplemented with glucose 30g/l. Color removal increased up to 5 days after 1 day lag time and leveled off. DyP activity appeared after 5 days and the MnP activity was significantly weak. The activities of LiP and Lac were not detected.

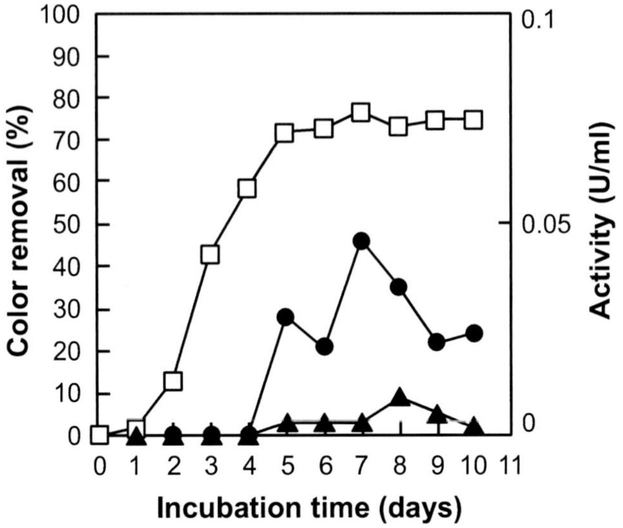

Fig. (69). Changes in color removal and enzyme activities when E-effluent was supplemented with glucose 30g/1 [216]. (□): color removal, (●): DyP activity, (▲): manganese peroxidase activity.

4.3. Discussion

Dec 1 exhibited decolorizing activity for commercial lignin on plate test, as shown in Fig. (**64**). The major colored substances of E-effluent were lignin and its related compounds, and thus, the decolorization of E-effluent is judged mainly due to the degradation of lignin by the Dec 1 cells.

As E-effluent contained no carbon sources available for Dec 1, neither color removal by Dec 1 nor growth of Dec 1 was observed. When glucose was supplemented, Dec 1 showed decolorization ability against the E-effluent, and at the glucose concentrations 20-30 g/1, color removal reached more than 70% after 4 day-cultivation. The increase in pH after complete consumption of glucose suggested that some by-products were converted from glucose by Dec 1, and were further consumed by Dec 1. The decreased color removal in accordance with the pH increase suggested that the color substances adsorbed to the cells detached from the cells. The desorption of the adsorbed colored substances is clearly shown in Table **33**. At 5 g/l glucose concentration, apparent decolorization was mainly due to the adsorption of colored substances to the cells of Dec 1, as shown in Table **33**. Increased glucose concentration decreased the amount of adsorbed colored substances to the Dec 1.

The molecular weight of the main colored substances of the E-effluent was estimated to be 5600 by gel filtration chromatography and the molecular weights of these substances were smaller compared to those of other natural polymers, such as polysaccharides and proteins. As the molecular weight of the E-effluent treated with Dec 1 changed from 5600 to less than 3000, the main structures of the degraded colored substances were relatively smaller compounds.

The decrease in AOX is an important factor in evaluating the degree of degradation of halogen-containing aromatic compounds. The decreased AOX content was reported to be 48%-65% and 23%-32% by an activated sludge method and an aerobic lagoon method, respectively [219, 220], and the contents of AOX decreased by 40% and 25% by different fungi strains KS-62 and *C. versicolor*, respectively [220]. The degradation degree of 43% AOX by Dec 1 was comparable to the results provided in previous reports.

LiP and MnP have been reported to be representative enzymes for lignin degradation and decolorization of kraft pulp bleaching effluent [220 - 222]. MnP was more effective than LiP for decolorizing kraft pulp bleaching effluent [217, 223, 224]. However, *Trametes versicolor* especially depended on laccase for decolorization of kraft pulp bleaching effluent [74]. Therefore, not only extracellular enzymes, like MnP and Lip, but also intracellular enzymes, can be important for the decolorization process.

In this study, the role of enzymes seemed to be different from the cases previously reported of the decolorization of E- effluent. DyP showed more than 70% color removal of the E-effluent, but MnP activity was low and LiP and Lac activities were not observed, as shown in Fig. (**69**). DyP, involved in decolorizing synthetic dyes, did not express to be a primary enzyme in decolorizing the E-effluent because the activity of DyP has been found to be delayed compared to the decolorizing timing of E-effluent by Dec 1.

Therefore, the decolorization activity of E-effluent by Dec 1 is speculated as follows: color removal by Dec 1 proceeds via two steps. In the first step, colored substances are adsorbed to the cells of Dec 1. The colored cells of Dec 1 increase gradually up to 3 to 4 days of culture. The second step involves the desorption and decomposition of the adsorbed colored substances to the cells intracellularly or extracellularly via the Dec 1 degradation of glucose. Glucose is essential for the growth of Dec 1 and for sufficient degradation of colored substances by Dec 1.

Fig. (**70**) shows changes in the colors of the kraft pulp bleaching effluents.

Fig. (70). Change in the colors of the kraft pulp bleaching effluents (left 2 flasks) compared to Dec 1 treated one (right) (prepared by the author).

5. DECOLORIZATION OF OXYGEN-DELIGNIFIED BLEACHING EFFLUENT (OBE) AND BIOBLEACHING OF OXYGEN-DELIGNIFIED KRAFT PULP (OKP) BY DEC 1

5.1. Methods

5.1.1. Strain

The details of the non-white-rot fungus Dec 1 have been described in previous chapters [1].

5.1.2. OBE and Oxygen-delignified Kraft Pulp (OKP)

The effluent produced from the oxygen extraction stage of hardwood pulp was named the Oxygen-delignified Bleaching Effluent (OBE). The pulp produced after the oxygen extraction stage of hard wood pulp was named the Oxygen-delignified Kraft Pulp (OKP). The initial brightness and kappa value of OKP have been previously determined as 47.8% and 12.4, respectively [225].

5.1.3. Medium

The Potato Dextrose (PD) medium was prepared using the method described in the previous section 4-1-1.

5.1.4. Decolorization of OBE by Dec 1

5.1.4.1. Decolorization

The OBE was centrifuged to remove insoluble particles, and glucose, ammonium tartrate, KH_2PO_4, $MgSO_4 .7H_2O$, and yeast extract were added to OBE. The two media, OGY 10 medium and OGY30 medium, depending on glucose concentrations of 10g and 30g, respectively, were prepared. Dec 1 spore suspension was inoculated into the media and cultivated at 30°C.

5.1.4.2. Color Removal

The color of the OBE was determined by the standard method of the Canadian Pulp and Paper Association [216]. The detail is provided in the previous section 4-1-5.

The color removal (%) was defined as the ratio of the Color Unit (CU) of the cell-free culture supernatant to that of the initial medium.

5.1.5. Biobleaching of OKP by Dec 1 Grown in OGY30 Medium

5.1.5.1. Cultivation of Dec 1

Dec 1 spore suspension was inoculated into the OGY30 medium and cultivated.

5.1.5.2. Biobleaching of OKP by Dec 1

Culture of Dec 1 grown in OGY30 medium was added to OKP suspension, including 2% (w/v) of pulp concentration, and shaken at 30°C for 15 days. Biobleaching was conducted using the method described previously [81]. On the other hand, 25% of pulp concentration was also prepared and biobleached by the same method as 2% pulp concentration to observe any difference in bleaching efficiency.

5.1.6. Biobleaching of OKP by the Culture after the Decolorization of OBE by Dec 1

5.1.6.1. Decolorization of OBE by Dec 1

The OBE decolorization by Dec 1 in the OGY30 medium has been described in a previous section.

5.1.6.2. Biobleaching of OKP by the Culture Filtrate

The OBE fluid prepared above was filtered through the gauze and by a membrane filter. The filtrate was mixed with 2 g of OKP containing 2% of pulp-concentration and shaken. As a control, a mixture of sterilized medium and 4 g of OKP was used.

5.1.7. Measurement of Brightness and the Kappa Value of Pulp

After the biobleaching of OKP, a sheet of OKP was prepared and the brightness of the sheet of OKP was determined with a spectrophotometric colorimeter according to JIS (Japanese Industrial Standard P 8209 and JIS P 8123). The kappa value of the sheet was measured according to the TAPPI standard [226].

5.1.8. Enzyme Assay

The activities of DyP, Manganese Peroxidase (MnP), Lignin Peroxidase (LiP), and Laccase (Lac) were determined using the same methods previously described.

5.2. Results

5.2.1. Decolorization of OBE by Dec 1

Fig (**71**) shows changes in color removal and the pH of OBE in two glucose media. Color removal reached 65% after 3 days of culture prepared in OGY 10 medium, but the pH increased when color removal decreased after 4 days, suggesting that the decolorization activity of Dec 1 stopped due to exhaustion of glucose. On the other hand, in the OGY30 medium, color removal increased similarly to that in the OGY 10 medium and reached 72% after 3 days of culture and leveled off. The continuous decline of pH, as shown in Fig. (**71**), was reflected by the gradual glucose consumption. In the medium of 0 g/1 of glucose, no color removal occurred. The color removal and enzyme activities of the culture supernatant are shown in Fig. (**72**). DyP and MnP activities appeared, but the activities of LiP and Lac were not observed.

Color removal by Dec 1 is supposed to proceed through the following two steps that are similar to those of E-effluent described in a previous section. The first step is the adsorption of colored substances to the cells of Dec 1. The second step is the desorption and decomposition of the adsorbed colored substances by the cells. After 4 days, DyP and Mn activities were activated, and the adsorbed colored substances were supposed to be decomposed by the enzymes.

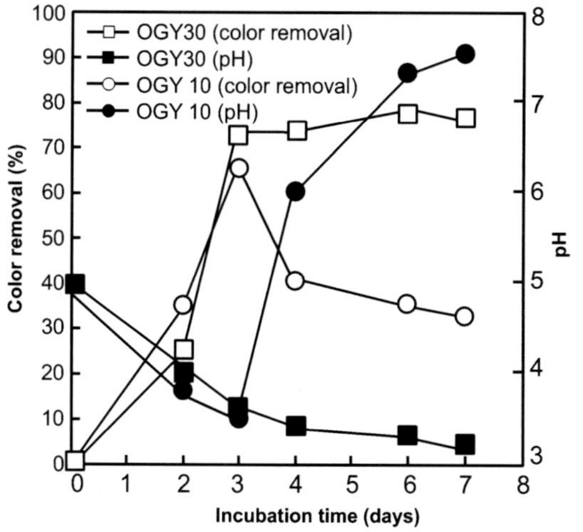

Fig. (71). Changes in color removal and the pH of OBE in two glucose media [225].

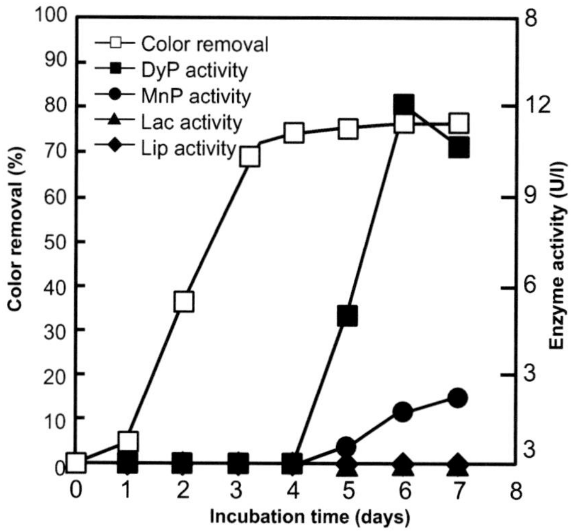

Fig. (72). Changes in color removal of OBE and enzyme activities of DyP, MnP, Lac, and Lip from Dec 1 culture grown in OGY30 medium [225].

5.2.2. Biobleaching of OKP by Dec 1 Cultured in OGY30 Medium

Fig. (**73**) shows changes in the biobleaching of OKP containing 2% pulp-concentration by Dec 1 culture grown in OGY30 medium. The brightness of OKP increased from 47.8% to 67% and the kappa value of OKP decreased from 12.4 to 4.2 points in 12 days.

When the pulp concentration of OKP increased to 25%, the brightness of OKP increased only from 47.8% to 52.1% and the kappa value of OKP decreased by 3 points during a 12-day incubation period. As oxygen is essential in the decolorization of various synthetic dyes by Dec 1 [1], oxygen concentration may be limited at 25% pulp concentration even if enough glucose remains. Fig. (**74**) shows changes in enzyme activities during bleaching of OKP at 2% pulp concentration. DyP and MnP activities appeared from the start of the experiment, but the activities of LiP and Lac were not observed.

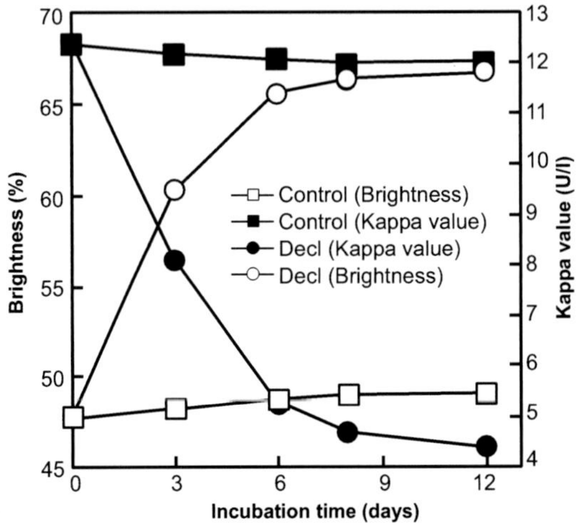

Fig. (73). Changes in the biobleaching of OKP at 2% pulp concentration by Dec 1 culture grown in OGY30 medium [225].

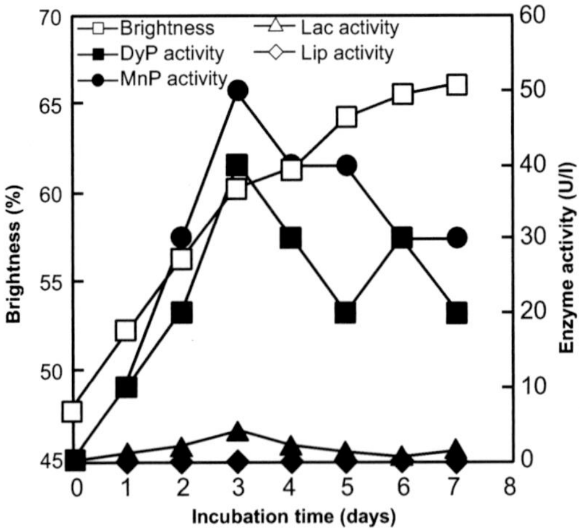

Fig. (74). Changes in enzyme activities during OKP bleaching at 2% pulp concentration by Dec 1 culture grown in OGY30 medium [225].

5.2.3. Biobleaching of OKP by the Culture Filtrate After the Decolorization of OBE by Dec 1

Fig. (**71**) shows that Dec 1 decolorized 70% of OBE in 3 days in OGY 30 medium. By using the cell-free culture after decolorization of OBE, as shown in Fig. (**71**), bleaching of OKP containing 2% pulp concentration was attempted.

Fig. (**75**) shows changes in enzyme activities during OKP bleaching using the cell-free culture fluid after the decolorization of OBE by Dec1. The brightness of OKP increased to 63% and DyP and MnP activities appeared during OKP biobleaching. On the other hand, the activities of LiP and Lac were not observed.

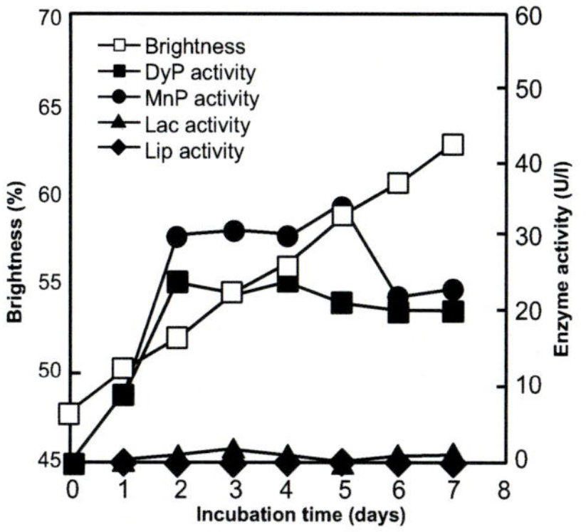

Fig. (75). Changes in enzyme activities during biobleaching of OKP using the cell-free culture fluid after the decolorization of OBE by Dec 1 [225].

Fig. (**76**) shows the kraft pulp before (left) and after (right) the bleaching treatment by Dec 1 culture.

Fig. (76). The kraft pulp before the bleaching treatment by Dec 1 culture (left) and afterward (right) (prepared by the author).

This showed the bleaching of OKP to be possible even using Dec 1 culture after decolorizing OBE.

CONCLUSION

In order to enhance decolorization by Dec 1, especially to achieve the highest decolorization degree of molasses, a jar fermenter with oxygen-enriched supply was effecive.

When Dec 1 decolorized molasses and RB5 dye simultaneously, the decolorization degree of molasses reached 87%, but the apparent activity of DyP produced in the molasses was significantly lower than that of purified DyP, suggesting the inhibition by molasses against DyP activity. When molasses was ultrafiltered into two fractions with different molecular weights, Dec 1 degraded both fractions significantly, indicating that Dec 1 decomposed colored substances with a wide range of molecular weights in molasses.

When Dec 1 cells were immobilized on polyurethane foam, decolorization of molasses, as well as DyP activity, maintained stably for more than eight weeks, indicating the advantage of immobilization and repeated batch operation in the decolorization of molasses and dyes.

When Dec 1 was applied for the decolorization of kraft pulp Oxygen-delignified Bleaching Effluent (OBE), the color removal ratio and the reduction degree of Absorbable Organic Halogens (AOX) reached 78% and 43%, respectively. Decolorization of OBE by Dec 1 was observed to proceed in two steps as follows: the first step involved the adsorption of colored substances to the cells of Dec 1 and the second step included desorption and decomposition of the adsorbed colored substances by the two enzymes, DyP and MnP, produced by Dec 1.

Dec 1 was also used to decolorize Oxygen-delignified Kraft Pulp (OKP), and more than 70% of the decolorization ratio was obtained. Biobleaching of OKP was possible not only by the Dec 1 culture fluid, but also by the cell-free culture broth prepared after decolorization of OBE by Dec 1.

REFERENCES

[1] Kim SJ, Ishikawa K, Hirai M, Shoda M. Characteristics of a newly isolated fungus, *Geotrichum candidum* Dec 1, which decolorizes various dyes. J Ferment Bioeng 1995; 79(6): 601-7. [http://dx.doi.org/10.1016/0922-338X(95)94755-G]

[2] Klein C. Production volume of synthetic dyes Japan 2012. Statista 2021; 2022: 27.

[3] Vaidya AA, Datye KV. Environmental pollution during chemical processing of synthetic fibers. Colourage 1982; 14: 3-10.

[4] Paszczynski A, Pasti-Grigsby MB, Goszczynski S, Crawford RL, Crawford DL. Mineralization of sulfonated azo dyes and sulfanilic acid by *Phanerochaete chrysosporium* and *Streptomyces chromofuscu*s. Appl Environ Microbiol 1992; 58(11): 3598-604. [http://dx.doi.org/10.1128/aem.58.11.3598-3604.1992] [PMID: 1482182]

[5] Spadaro JT, Gold MH, Renganathan V. Degradation of azo dyes by the lignin-degrading fungus *Phanerochaete chrysosporium.* Appl Environ Microbiol 1992; 58(8): 2397-401. [http://dx.doi.org/10.1128/aem.58.8.2397-2401.1992] [PMID: 1514787]

[6] Platt M, Hadar Y, Chet I. The decolorization of the polymeric dye Poly-Blue (polyvinalamine sulfonate-anthroquinone) by lignin degrading fungi. Appl Microbiol Biotechnol 1985; 21(6): 394-6. [http://dx.doi.org/10.1007/BF00249987]

[7] Watanabe Y, Sugi R, Tanaka Y, Hayashida S. Enzymatic decolorization of melanoidin by *Coriolus* sp. No. 20. Agric Biol Chem 1982; 46(6): 1623-30. [http://dx.doi.org/10.1271/bbb1961.46.1623]

[8] Kang SO, Shin KS, Han YH, Youn HD, Hah YC. Purification and characterisation of an extracellular peroxidase from white-rot fungus *Pleurotus ostreatus.* Biochim Biophys Acta Protein Struct Mol Enzymol 1993; 1163(2): 158-64. [http://dx.doi.org/10.1016/0167-4838(93)90177-S] [PMID: 8387825]

[9] Ollikka P, Alhonmäki K, Leppänen VM, Glumoff T, Raijola T, Suominen I. Decolorization of azo, triphenyl methane, heterocyclic, and polymeric dyes by lignin peroxidase isozymes from *Phanerochaete chrysosporium.* Appl Environ Microbiol 1993; 59(12): 4010-6. [http://dx.doi.org/10.1128/aem.59.12.4010-4016.1993] [PMID: 16349103]

[10] Goszczynski S, Paszczynski A, Pasti-Grigsby MB, Crawford RL, Crawford DL. New pathway for degradation of sulfonated azo dyes by microbial peroxidases of *Phanerochaete chrysosporium* and *Streptomyces chromofuscus.* J Bacteriol 1994; 176(5): 1339-47. [http://dx.doi.org/10.1128/jb.176.5.1339-1347.1994] [PMID: 8113173]

[11] Chivukula M, Renganathan V. Phenolic azo dye oxidation by laccase from *Pyricularia oryzae.* Appl Environ Microbiol 1995; 61(12): 4374-7. [http://dx.doi.org/10.1128/aem.61.12.4374-4377.1995] [PMID: 16535191]

[12] Chivukula M, Spadaro JT, Renganathan V. Lignin peroxidase-catalyzed oxidation of sulfonated azo dyes generates novel sulfophenyl hydroperoxides. Biochemistry 1995; 34(23): 7765-72. [http://dx.doi.org/10.1021/bi00023a024] [PMID: 7779823]

[13] Vyas BR, Molitoris HP. Involvement of an extracellular H_2O_2-dependent ligninolytic activity of the white rot fungus *Pleurotus ostreatus* in the decolorization of Remazol brilliant blue R. Appl Environ Microbiol 1995; 61(11): 3919-27. [http://dx.doi.org/10.1128/aem.61.11.3919-3927.1995] [PMID: 8526504]

[14] Shin K, Oh I, Kim C. Production and purification of Remazol brilliant blue R decolorizing peroxidase from the culture filtrate of *Pleurotus ostreatus.* Appl Environ Microbiol 1997; 63(5): 1744-8. [http://dx.doi.org/10.1128/aem.63.5.1744-1748.1997] [PMID: 16535590]

Makoto Shoda
All rights reserved-© 2024 Bentham Science Publishers

[15] Heinfling A, Martínez MJ, Martínez AT, Bergbauer M, Szewzyk U. Purification and characterization of peroxidases from the dye-decolorizing fungus *Bjerkandera adusta.* FEMS Microbiol Lett 1998; 165(1): 43-50.
[http://dx.doi.org/10.1111/j.1574-6968.1998.tb13125.x] [PMID: 9711838]

[16] Young L, Yu J. Ligninase-catalysed decolorization of synthetic dyes. Water Res 1997; 31(5): 1187-93.
[http://dx.doi.org/10.1016/S0043-1354(96)00380-6]

[17] Camarero S, Böckle B, Martínez MJ, Martínez AT. Manganese-mediated lignin degradation by *Pleurotus pulmonarius.* Appl Environ Microbiol 1996; 62(3): 1070-2.
[http://dx.doi.org/10.1128/aem.62.3.1070-1072.1996] [PMID: 16535257]

[18] Heinfling A, Martínez MJ, Martínez AT, Bergbauer M, Szewzyk U. Transformation of industrial dyes by manganese peroxidases from *Bjerkandera adusta* and *Pleurotus eryngii* in a manganese-independent reaction. Appl Environ Microbiol 1998; 64(8): 2788-93.
[http://dx.doi.org/10.1128/AEM.64.8.2788-2793.1998] [PMID: 9687431]

[19] Ruiz-Dueñas FJ, Camarero S, Pérez-Boada M, Martínez MJ, Martínez AT. A new versatile peroxidase from *Pleurotus.* Biochem Soc Trans 2001; 29(2): 116-22.
[http://dx.doi.org/10.1042/bst0290116] [PMID: 11356138]

[20] Kim SJ, Shoda M. Purification and characterization of a novel peroxidase from *Geotrichum candidum* dec 1 involved in decolorization of dyes. Appl Environ Microbiol 1999; 65(3): 1029-35.
[http://dx.doi.org/10.1128/AEM.65.3.1029-1035.1999] [PMID: 10049859]

[21] Kim SJ, Shoda M. Decolorization of molasses and a dye by a newly isolated strain of the fungus *Geotrichum candidum* Dec 1. Biotechnol Bioeng 1999; 62(1): 114-9.
[http://dx.doi.org/10.1002/(SICI)1097-0290(19990105)62:1<114::AID-BIT13>3.0.CO;2-T] [PMID: 10099519]

[22] Kim SJ, Suzuki N, Uematsu Y, Shoda M. Characterization of aryl alcohol oxidase produced by dye-decolorizing fungus *Geotrichum candidum* Decl. J Biosci Bioeng 2001; 91(2): 166-72.
[http://dx.doi.org/10.1016/S1389-1723(01)80060-8] [PMID: 16232969]

[23] Sugano Y, Matsushima Y, Shoda M. Complete decolorization of the anthraquinone dye Reactive blue 5 by the concerted action of two peroxidases from *Thanatephorus cucumeris* Dec 1. Appl Microbiol Biotechnol 2006; 73(4): 862-71.
[http://dx.doi.org/10.1007/s00253-006-0545-9] [PMID: 16944133]

[24] Sugano Y, Muramatsu R, Ichiyanagi A, Sato T, Shoda M. DyP, a unique dye-decolorizing peroxidase, represents a novel heme peroxidase family: ASP171 replaces the distal histidine of classical peroxidases. J Biol Chem 2007; 282(50): 36652-8.
[http://dx.doi.org/10.1074/jbc.M706996200] [PMID: 17928290]

[25] Welinder KG. Superfamily of plant, fungal and bacterial peroxidases. Curr Opin Struct Biol 1992; 2(3): 388-93.
[http://dx.doi.org/10.1016/0959-440X(92)90230-5]

[26] Kaput J, Goltz S, Blobel G. Nucleotide sequence of the yeast nuclear gene for cytochrome c peroxidase precursor. Functional implications of the pre sequence for protein transport into mitochondria. J Biol Chem 1982; 257(24): 15054-8.
[http://dx.doi.org/10.1016/S0021-9258(18)33392-1] [PMID: 6294090]

[27] Welinder KG. Bacterial catalase-peroxidases are gene duplicated members of the plant peroxidase superfamily. Biochim Biophys Acta Protein Struct Mol Enzymol 1991; 1080(3): 215-20.
[http://dx.doi.org/10.1016/0167-4838(91)90004-J] [PMID: 1954228]

[28] Pribnow D, Mayfield MB, Nipper VJ, Brown JA, Gold MH. Characterization of a cDNA encoding a manganese peroxidase, from the lignin-degrading basidiomycete *Phanerochaete chrysosporium.* J Biol Chem 1989; 264(9): 5036-40.
[http://dx.doi.org/10.1016/S0021-9258(18)83695-X] [PMID: 2925681]

[29] Ritch T G, Nipper V J, Akireswaran L, Smith A J, Pribnow D J, Gold MH. Lignin peroxidase from the basidiomycete *Phanerochaete chrysosporwm* is synthesized as a preproenzyme. Gene 1991; 107: 9-126.

[30] Sawai-Hatanaka H, Ashikari T, Tanaka Y, *et al.* Cloning, sequencing, and heterologous expression of a gene coding for *Arthromyces ramosus* peroxidase. Biosci Biotechnol Biochem 1995; 59(7): 1221-8. [http://dx.doi.org/10.1271/bbb.59.1221] [PMID: 7670182]

[31] Welinder KG. Covalent structure of the glycoprotein horseradish peroxidase (EC 1.11.1.7). FEBS Lett 1976; 72(1): 19-23. [http://dx.doi.org/10.1016/0014-5793(76)80804-6] [PMID: 1001465]

[32] Welinder KG, Mazza G. Amino-acid sequences of heme-linked, histidine-containing peptides of five peroxidases from horseradish and turnip. Eur J Biochem 1977; 73(2): 353-8. [http://dx.doi.org/10.1111/j.1432-1033.1977.tb11325.x] [PMID: 849740]

[33] Barbesgaard P, Heldt-Hansen H, Diderichsen B. On the safety of *Aspergillus oryzae:* a review. Appl Microbiol Biotechnol 1992; 36(5): 569-72. [http://dx.doi.org/10.1007/BF00183230] [PMID: 1368061]

[34] Linko S. Continuous production of lignin peroxidase by immobilized *Phanerochaete chrysosporium* in a pilot scale bioreactor. J Biotechnol 1988; 8(2): 163-70. [http://dx.doi.org/10.1016/0168-1656(88)90078-8]

[35] Moreira MT, Palma C, Feijoo G, Lema JM. Strategies for the continuous production of ligninolytic enzymes in fixed and fluidised bed bioreactors. J Biotechnol 1998; 66(1): 27-39. [http://dx.doi.org/10.1016/S0168-1656(98)00154-0]

[36] Shuler ML, Karagi F. Bioprocess engineering: basic concepts. Englewood Cliffs: Prentice Hall 1992; pp. 148-234.

[37] Feijoo G, Dosoretz C, Lema JM. Production of lignin peroxidase from *Phanerochaete chrysosporium* in a packed bed bioreactor with recycling. Biotechnol Tech 1994; 8(5): 363-8. [http://dx.doi.org/10.1007/BF02428982]

[38] Feijoo G, Dosoretz C, Lema JM, Lema JM. Production of lignin peroxidase by *Phanerochaete chrysosporium* in a packed bed bioreactor operated in semi-continuous mode. J Biotechnol 1995; 42(3): 247-53. [http://dx.doi.org/10.1016/0168-1656(95)00085-5]

[39] Chang HN, Yoo IK, Kim BS. High density cell culture by membrane-based cell recycle. Biotechnol Adv 1994; 12(3): 467-87. [http://dx.doi.org/10.1016/0734-9750(94)90020-5] [PMID: 14548467]

[40] Ozturk SS. Engineering challenges in high density cell culture systems. Cytotechnology 1996; 22(1-3): 3-16. [http://dx.doi.org/10.1007/BF00353919] [PMID: 22358910]

[41] Gomyo T, Kato H, Udaka K, Horikoshi M, Fujimaki M. Chemical studies on melanoidins. Part III. Effects of heating on chemical properties of melanoidin prepared from glycine-xylose system. Agric Biol Chem 1972; 36: 125-32.

[42] Sirianuntapiboon S, Somchai P, Sihanonth P, Attharnmpunna P, Ohmomo S. Microbial decolorization of molasse wastewater by *Mycelia sterilia* D90. Agric Biol Chem 1988; 52:393-398. [http://dx.doi.org/10.1271/bbb1961.52.393]

[43] Miranda M P, Benito GG, Cristobal NS, Nieto CH. Color elimination from molasses wastewater by *Aspergl llus niger*. Biores Technol 1996; 57: 229-235. [http://dx.doi.org/10.1016/S0960-8524(96)00048-X]

[44] Fahy V, Fitzgibbon FJ, McMullan G, Singh D, Marchant R. Decolourisation of molasses spent wash by *Phanerochaete chrysosporium*. Biotechnol Lett 1997; 19: 97-99.

[http://dx.doi.org/10.1023/A:1018335608459]

[45] Lee TH, Aoki H, Sugano Y, Shoda M. Effect of molasses on the production and activity of dye-decolorizing peroxidase from *Geotrichum candidum* Dec l. J Ferment Bioeng 2000; 89: 545-549. [http://dx.doi.org/10.1016/S1389-1723(00)80054-7] [PMID: 16232795]

[46] Kim S J, Shoda M. Decolorization of molasses by a new isolate of *Geotrichum candidum* in a jar fermentor. Biotechnol Tech 1998; 12: 497-499. [http://dx.doi.org/10.1023/A:1008824119174]

[47] Kim S J, Shoda M. Batch decolorization of molasses by suspended and immobilized fungus of *Geotrichum candidum* Dec 1. J Biosci Bioeng 1999; 88: 586-589. [http://dx.doi.org/10.1016/S1389-1723(00)87682-3] [PMID: 16232668]

[48] Zhang J, Greasham R. Chemical defined media for commercial fermentations. Appl Microbial Biotechnol 1999; 51: 407-421. [http://dx.doi.org/10.1007/s002530051411]

[49] Tsuchiya K, Nagashima T, Yamamoto Y, Gomi K, Kitamoto K, Kumagai C, Tamura G. High level secretion of chymosin using a glucoamylase-prochymosin fusion gene in *Aspergillus oryzae*. Biosci Biotechnol Biochem 1994; 58: 895-899. [http://dx.doi.org/10.1271/bbb.58.895] [PMID: 7764977]

[50] Shoda M. Biocontrol of Plant Disease by *Bacillus subtilis*. CRC Press, Taylor & Francis Group, Boca Raton, FL, U.S.A. 2020; pp.181-207.

[51] Oda K, Kakizono D, Yamada O, Iefuji H, Akita O, Iwashita K. Proteomic analysis of extracellular proteins from *Aspergillus oryzae* grown under submerged and solid-state culture conditions. Appl Environ Microbiol 2006; 72(5): 3448-57. [http://dx.doi.org/10.1128/AEM.72.5.3448-3457.2006] [PMID: 16672490]

[52] Ashokkumar B, Kayalvizhi N, Gunasekaran P. Optimization of media for β-fructofuranosidase production by *Aspergillus niger* in submerged and solid state fermentation. Process Biochem 2001; 37 (4): 331-8. [http://dx.doi.org/10.1016/S0032-9592(01)00204-7]

[53] Asther M, Haon M, Roussos S, *et al.* Feruloyl esterase from *Aspergillus niger*. Process Biochem 2002; 38(5): 685-91. [http://dx.doi.org/10.1016/S0032-9592(02)00196-6]

[54] Mahadik ND, Puntambekar US, Bastawde KB, Khire JM, Gokhale DV. Production of acidic lipase by *Aspergillus niger* in solid state fermentation. Process Biochem 2002; 38(5): 715-21. [http://dx.doi.org/10.1016/S0032-9592(02)00194-2]

[55] Yasuhara A, Ogawa A, Tanaka T, Sakiyama T, Nakanishi K. Production of neutral protease from *Aspergillus oryzae* by a novel cultivation method on a microporous membrane. Biotechnol Tech 1994; 8(4): 249-54. [http://dx.doi.org/10.1007/BF00155416]

[56] Yan L, Boyd KG, Adams DR, Burgess JG. Biofilm-specific cross-species induction of antimicrobial compounds in bacilli. Appl Environ Microbiol 2003; 69(7): 3719-27. [http://dx.doi.org/10.1128/AEM.69.7.3719-3727.2003] [PMID: 12839737]

[57] Ogawa A, Yasuhara A, Tanaka T, Sakiyama T, Nakanishi K. Production of neutral protease by membrane-surface liquid culture of *Aspergillus oryzae* IAM2704. J Ferment Bioeng 1995; 80(1): 35-40. [http://dx.doi.org/10.1016/0922-338X(95)98173-I]

[58] Tanaka T, Eguchi S, Aoki T, *et al.* Production of laccase by membrane-surface liquid culture of *Trametes versicolor* using a poly(L-lactic acid) membrane. Biochem Eng 2007; 33: 188-91.

[59] Nicell JA, Al-Kassim L, Bewtra JK, Taylor KE. Wastewater treatment by enzyme catalysed polymerization and precipitation. Biodegrad Abstr 1993; 7: 1-8.

[60] Karam J, Nicell JA. Potential application of enzymes in waste treatment. J Chem Technol Biotechnol 1997; 69(2): 141-53.
[http://dx.doi.org/10.1002/(SICI)1097-4660(199706)69:2<141::AID-JCTB694>3.0.CO;2-U]

[61] Conesa A, Punt PJ, van den Hondel CAMJJ. Fungal peroxidases: molecular aspects and applications. J Biotechnol 2002; 93(2): 143-58.
[http://dx.doi.org/10.1016/S0168-1656(01)00394-7] [PMID: 11738721]

[62] Shakeri M, Shoda M. Change in turnover capacity of crude recombinant dye-decolorizing peroxidase (rDyP) in batch and fed-batch decolorization of Remazol Brilliant Blue R. Appl Microbiol Biotechnol 2007; 76(4): 919-26.
[http://dx.doi.org/10.1007/s00253-007-1042-5] [PMID: 17569040]

[63] Saijo S, Sato T, Tanaka N, Ichiyanagi A, Sugano Y, Shoda M. Precipitation diagram and optimization of crystallization conditions at low ionic strength for deglycosylated dye-decolorizing peroxidase from a basidiomycete. Acta Crystallogr Sect F Struct Biol Cryst Commun 2005; 61(8): 729-32.
[http://dx.doi.org/10.1107/S1744309105019469] [PMID: 16511141]

[64] Takahashi H, Li B, Sasaki T, Miyazaki C, Kajino T, Inagaki S. Catalytic activity in organic solvents and stability of immobilized enzyme depend on the pore size and surface characteristic of mesoporous silica. Chem Mater 2000; 12(11): 3301-5.
[http://dx.doi.org/10.1021/cm000487a]

[65] Washmon-Kriel L, Jimenez VL, Balkus KJ Jr. Cytochrome c immobilization into mesoporous molecular sieves. J Mol Catal, B Enzym 2000; 10(5): 453-69.
[http://dx.doi.org/10.1016/S1381-1177(99)00123-X]

[66] Yiu HHP, Wright PA. Enzymes supported on ordered mesoporous solids: a special case of an inorganic–organic hybrid. J Mater Chem 2005; 15(35-36): 3690-700.
[http://dx.doi.org/10.1039/b506090g]

[67] Sugano Y, Matsuo C, Shoda M. Efficient production of a heterologous peroxidase, DyP from *Geotrichum candidum* Dec 1, on solid-state culture of *Aspergillus oryzae* RD005. J Biosci Bioeng 2001; 92(6): 594-7.
[http://dx.doi.org/10.1016/S1389-1723(01)80323-6] [PMID: 16233153]

[68] Aitken MD, Heck PE. Turnover capacity of *Coprinus cinereus* peroxidase for phenol and monosubstituted phenols. Biotechnol Prog 1998; 14(3): 487-92.
[http://dx.doi.org/10.1021/bp980034z] [PMID: 9622531]

[69] Prouty AL. Bench-scale development and evaluation of a fungal bioreactor for color removal from bleach effluents. Appl Microbiol Biotechnol 1990; 32(4): 490-3.
[http://dx.doi.org/10.1007/BF00903789]

[70] Smeds A, Holmbom B, Pettersson C. Chemical stability of chlorinated components in pulp bleaching liquors. Chemosphere 1994; 28(5): 881-95.
[http://dx.doi.org/10.1016/0045-6535(94)90005-1]

[71] Eilbeck WV, Mattock G. Chemical and adsorptive precipitation and flocculation.Chemical processes in wastewater treatment. New York: Wiley 1987; pp. 32-293.

[72] Livernoche D, Jurasek L, Desrochers M, Dorica J, Veliky IA. Removal of color from kraft mill wastewaters with cultures of white-rot fungi and with immobilized mycelium of *Coriolus versicolor.* Biotechnol Bioeng 1983; 25(8): 2055-65.
[http://dx.doi.org/10.1002/bit.260250814] [PMID: 18551550]

[73] Royer G, Livernoche D, Desrochers M, Jurasek L, Rouleau D, Mayer RC. Decolorization of kraft mill effluent: kinetics of a continuous process using immobilized *Coriolus versicolor.* Biotechnol Lett 1983; 5: 321-6.

[74] Archibald F, Roy B. Production of manganic chelates by laccase from the lignin-degrading fungus *Trametes (Coriolus) versicolor.* Appl Environ Microbiol 1992; 58(5): 1496-9.

[http://dx.doi.org/10.1128/aem.58.5.1496-1499.1992] [PMID: 1622216]

[75] Jaspers CJ, Jimenez G, Penninckx MJ. Evidence for a role of manganese peroxidase in the decolorization of Kraft pulp bleach plant effluent by *Phanerochaete chrysosporium*: Effects of initial culture conditions on enzyme production. J Biotechnol 1994; 37(3): 229-34. [http://dx.doi.org/10.1016/0168-1656(94)90130-9]

[76] Kringstad KP, Lindström K. Spent liquors from pulp bleaching. Environ Sci Technol 1984; 18(8): 236A-48A. [http://dx.doi.org/10.1021/es00126a714] [PMID: 22663368]

[77] Moreira MT, Feijoo G, Sierra-Alvarez R, Lema J, Field JA. Biobleaching of oxygen delignified kraft pulp by several white rot fungal strains. J Biotechnol 1997; 53(2-3): 237-51. [http://dx.doi.org/10.1016/S0168-1656(97)01676-3]

[78] Reid ID, Paice MG. Effect of residual lignin type and amount on bleaching of kraft pulp by *Trametes versicolor.* Appl Environ Microbiol 1994; 60(5): 1395-400. [http://dx.doi.org/10.1128/aem.60.5.1395-1400.1994] [PMID: 16349246]

[79] Murata S, Kondo R, Sakai K, Kashino Y, Nishida T, Takahara Y. Chlorine free bleaching process of kraft pulp using treatment with the fungus IZU-154. Tappi J 1992; 75: 91-4.

[80] Moreira MT, Feijoo G, Sierra-Alvarez R, Lema J, Field JA. Manganese is not required for biobleaching of oxygen-delignified kraft pulp by the white rot fungus *Bjerkandera* sp. strain BOS55. Appl Environ Microbiol 1997; 63(5): 1749-55. [http://dx.doi.org/10.1128/aem.63.5.1749-1755.1997] [PMID: 16535591]

[81] Iimori T, Kaneko R, Yoshikawa H, Machida M, Yoshioka H, Murakami K. Screening of pulp-bleaching fungi and bleaching activity of newly isolated fungus SKB-1152. Mokuzai Gakkaishi, (in Japanese) 1994; 40: 733-7.

[82] Kamilari E, Stanton C, Reen FJ, Ross RP. Uncovering the Biotechnological Importance of *Geotrichum candidum.* Foods 2023; 12(6): 1124. [http://dx.doi.org/10.3390/foods12061124] [PMID: 36981051]

[83] Jakovljević VD, Vrvić MM. Potential of pure and mixed cultures of *Cladosporium cladosporioides* and *Geotrichum candidum* for application in bioremediation and detergent industry. Saudi J Biol Sci 2018; 25(3): 529-36. [http://dx.doi.org/10.1016/j.sjbs.2016.01.020] [PMID: 29686515]

[84] Tian J-H, Pourcher A-M, Klingelschmitt F, Le Roux S, Peu P. Class P dye-decolorizing peroxidase gene: Degenerated primers design and phylogenetic analysis. J Microbiol Methods 2016; 130: 148-53. [http://dx.doi.org/10.1016/j.mimet.2016.09.016] [PMID: 27686379]

[85] Falade AO, Mabinya LV, Okoh AI, *et al.* Biochemical and molecular characterization of a novel dye-decolourizing peroxidase from *Raoultella ornithinolytica* OKOH-1. Int J Biol Macromol 2019; 121: 454-62. [http://dx.doi.org/10.1016/j.ijbiomac.2018.10.045] [PMID: 30316770]

[86] Guo G, Tian F, Zhao Y, *et al.* Aerobic decolorization and detoxification of Acid Scarlet GR by a newly isolated salt-tolerant yeast strain *Galactomyces geotrichum* GG. Int Biodeterior Biodegradation 2019; 145: 104818. [http://dx.doi.org/10.1016/j.ibiod.2019.104818]

[87] Chen C, Shrestha R, Jia K, *et al.* Characterization of Dye-decolorizing Peroxidase (DyP) from *Thermomonospora curvata* Reveals Unique Catalytic Properties of A-type DyPs. J Biol Chem 2015; 290(38): 23447-63. [http://dx.doi.org/10.1074/jbc.M115.658807] [PMID: 26205819]

[88] Musengi A, Durrell K, Prins A, *et al.* Production and characterisation of a novel actinobacterial DyP-type peroxidase and its application in coupling of phenolic monomers. Enzyme Microb Technol 2020; 141: 109654.

[http://dx.doi.org/10.1016/j.enzmictec.2020.109654] [PMID: 33051013]

[89] Linde D, Ayuso-Fernández I, Ruiz-Dueñas FJ, Martínez AT. Different fungal peroxidases oxidize nitrophenols at a surface catalytic tryptophan. Arch Biochem Biophys 2019; 668: 23-8. [http://dx.doi.org/10.1016/j.abb.2019.05.010] [PMID: 31095936]

[90] Uchida T, Omura I, Umetsu S, Ishimori K. Radical transfer but not heme distal residues is essential for pH dependence of dye-decolorizing activity of peroxidase from *Vibrio cholerae.* J Inorg Biochem 2021; 219: 111422. [http://dx.doi.org/10.1016/j.jinorgbio.2021.111422] [PMID: 33756393]

[91] Rath S, Paul M, Thatoi H. Molecular modeling, docking and dynamic studies of fungal DyPs to determine substrate specificity for an efficient lignin biodegradation towards bioethanol production. Bioresour Technol Rep 2022; 18: 101036. [http://dx.doi.org/10.1016/j.biteb.2022.101036]

[92] Bilal M, Asgher M, Iqbal HMN, Hu H, Wang W, Zhang X. Bio-catalytic performance and dye-based industrial pollutants degradation potential of agarose-immobilized MnP using a Packed Bed Reactor System. Int J Biol Macromol 2017; 102: 582-90. [http://dx.doi.org/10.1016/j.ijbiomac.2017.04.065] [PMID: 28431941]

[93] Yang HY, Liu J, Wang YX, *et al.* Bioelectrochemical decolorization of a reactive diazo dye: Kinetics, optimization with a response surface methodology, and proposed degradation pathway. Bioelectrochemistry 2019; 128: 9-16. [http://dx.doi.org/10.1016/j.bioelechem.2019.02.008] [PMID: 30884361]

[94] Dhankhar P, Dalal V, Singh N, Gurjar BR. Bioremediation of synthetic dyes: Dye decolorizing peroxidases (DyPs).Removal of Toxic Pollutants Through Microbiological and Tertiary Treatment. Elsevier 2020; pp. 453-86. [http://dx.doi.org/10.1016/B978-0-12-821014-7.00018-6]

[95] Akansha K, Kaur T, Yadav A, *et al.* Microbe-mediated remediation of dyes: Current status and future challenges. J Appl Biol Biotechnol 2023; 11: 1-23.

[96] Advances in eco-friendly and sustainable technologies for the treatment of textile wastewater. Curr Develop Bioengin Biotech. Ashok Pandey, Sanjay P. Govindwar SP, Mayur B, Kurade MB, Byong-Hun Jeon BH, Eds. Elsevier 2023.

[97] Pasti MB, Crawford DL. Relationships between the abilities of streptomycetes to decolorize three anthron-type dyes and to degrade lignocellulose. Can J Microbiol 1991; 37(12): 902-7. [http://dx.doi.org/10.1139/m91-156]

[98] Glenn JK, Gold MH. Decolorization of several polymeric dyes by the lignin-degrading basidiomycete *Phanerochaete chrysosporium.* Appl Environ Microbiol 1983; 45(6): 1741-7. [http://dx.doi.org/10.1128/aem.45.6.1741-1747.1983] [PMID: 16346307]

[99] Cripps C, Bumpus JA, Aust SD. Biodegradation of azo and heterocyclic dyes by *Phanerochaete chrysosporium.* Appl Environ Microbiol 1990; 56(4): 1114-8. [http://dx.doi.org/10.1128/aem.56.4.1114-1118.1990] [PMID: 2339873]

[100] Pasti-Grigsby MB, Paszczynski A, Goszczynski S, Crawford DL, Crawford RL. Influence of aromatic substitution patterns on azo dye degradability by *Streptomyces* spp. and *Phanerochaete chrysosporium.* Appl Environ Microbiol 1992; 58(11): 3605-13. [http://dx.doi.org/10.1128/aem.58.11.3605-3613.1992] [PMID: 1482183]

[101] Henderson MEK. The metabolism of aromatic compounds related to lignin by some hyphomycetes and yeast-like fungi of soil. J Gen Microbiol 1961; 26(1): 155-65. [http://dx.doi.org/10.1099/00221287-26-1-155] [PMID: 13906407]

[102] Tubaki K. Studies on a slime-forming fungus in polluted water. Trans Myc Soc Japan 1962; 3: 29-35.

[103] Klibanov AM, Morris ED. Horseradish peroxidase for the removal of carcinogenic aromatic amines from water. Enzyme Microb Technol 1981; 3(2): 119-22.

[http://dx.doi.org/10.1016/0141-0229(81)90069-7]

[104] Nakamoto S, Machida N. Phenol removal from aqueous solutions by peroxidase-catalyzed reaction using additives. Water Res 1992; 26(1): 49-54.
[http://dx.doi.org/10.1016/0043-1354(92)90110-P]

[105] Tien M, Kirk TK. Lignin-degrading enzyme from *Phanerochaete chrysosporium* : Purification, characterization, and catalytic properties of a unique H_2O_2 -requiring oxygenase. Proc Natl Acad Sci USA 1984; 81(8): 2280-4.
[http://dx.doi.org/10.1073/pnas.81.8.2280] [PMID: 16593451]

[106] Fiechter A. Function and synthesis of enzymes involved in lignin degradation. J Biotechnol 1993; 30(1): 49-55.
[http://dx.doi.org/10.1016/0168-1656(93)90026-J]

[107] Matsubara M, Suzuki J, Deguchi T, Miura M, Kitaoka Y. Characterization of manganese peroxidases from the hyperlignolytic fungus IZU-154. Appl Environ Microbiol 1996; 62(11): 4066-72.
[http://dx.doi.org/10.1128/aem.62.11.4066-4072.1996] [PMID: 8899997]

[108] Shannon LM, Kay E, Lew JY. Peroxidase isozymes from horseradish roots. I. Isolation and physical properties. J Biol Chem 1966; 241(9): 2166-72.
[http://dx.doi.org/10.1016/S0021-9258(18)96680-9] [PMID: 5946638]

[109] Paul KG, Stigbrand T, Nimmich W, Rönnquist O, Werner P-E. Four isoperoxidases from horse radish root. Acta Chem Scand 1970; 24: 3607-17.
[http://dx.doi.org/10.3891/acta.chem.scand.24-3607]

[110] Wang SS, Dimarco GR. Isolation and characterization of the native, thermally inactivated and regenerated horseradish peroxidase isozymes. J Food Sci 1972; 37(4): 574-8.
[http://dx.doi.org/10.1111/j.1365-2621.1972.tb02696.x]

[111] Glumoff T, Harvey PJ, Molinari S, *et al.* Lignin peroxidase from *Phanerochaete chrysosporium.* Eur J Biochem 1990; 187(3): 515-20.
[http://dx.doi.org/10.1111/j.1432-1033.1990.tb15333.x] [PMID: 2303054]

[112] Schmidt B, Heimgartner U, Kozulić B, Leisola MSA. Lignin peroxidases are oligomannose type glycoproteins. J Biotechnol 1990; 13(2-3): 223-8.
[http://dx.doi.org/10.1016/0168-1656(90)90107-M]

[113] Maehly AC. Plant peroxidase. In: S. Colowick, N.O. Kaplan, Eds. Methods in enzymology. New York, N.Y: Academic Press, Inc. 1955; pp. 801-3.

[114] Wariishi H, Gold MH. Lignin peroxidase compound III. FEBS Lett 1989; 243(2): 165-8.
[http://dx.doi.org/10.1016/0014-5793(89)80122-X]

[115] Valli K, Wariishi H, Gold MH. Oxidation of monomethoxylated aromatic compounds by lignin peroxidase: role of veratryl alcohol in lignin biodegradation. Biochemistry 1990; 29(37): 8535-9.
[http://dx.doi.org/10.1021/bi00489a005] [PMID: 2271536]

[116] Renganathan V, Gold MH. Spectral characterization of the oxidized states of lignin peroxidase, an extracellular heme enzyme from the white rot basidiomycete *Phanerochaete chrysosporium.* Biochemistry 1986; 25(7): 1626-31.
[http://dx.doi.org/10.1021/bi00355a027]

[117] Lundell T, Wever R, Floris R, *et al.* Lignin peroxidase L3 from *Phlebia radiata.* Pre-steady-state and steady-state studies with veratryl alcohol and a non-phenolic lignin model compound 1-(3, 4-dimethoxyphenyl)-2-(2-methoxyphenoxy)propane-1,3-diol. Eur J Biochem 1993; 211(3): 391-402.
[http://dx.doi.org/10.1111/j.1432-1033.1993.tb17562.x] [PMID: 8436103]

[118] Wariishi H, Valli K, Gold MH. Manganese(II) oxidation by manganese peroxidase from the basidiomycete *Phanerochaete chrysosporium.* Kinetic mechanism and role of chelators. J Biol Chem 1992; 267(33): 23688-95.
[http://dx.doi.org/10.1016/S0021-9258(18)35893-9] [PMID: 1429709]

[119] Kuan IC, Johnson KA, Tien M. Kinetic analysis of manganese peroxidase. The reaction with manganese complexes. J Biol Chem 1993; 268(27): 20064-70. [http://dx.doi.org/10.1016/S0021-9258(20)80694-2] [PMID: 8376363]

[120] Faison BD, Kirk TK. Factors involved in the regulation of a ligninase activity in *Phanerochaete chrysosporium.* Appl Environ Microbiol 1985; 49(2): 299-304. [http://dx.doi.org/10.1128/aem.49.2.299-304.1985] [PMID: 16346716]

[121] Kirk TK, Schultz E, Connors WJ, Lorenz LF, Zeikus JG. Influence of culture parameters on lignin metabolism by *Phanerochaete chrysosporium.* Arch Microbiol 1978; 117(3): 277-85. [http://dx.doi.org/10.1007/BF00738547]

[122] Aoshima I, Tozawa Y, Ohmomo S, Ueda K. Production of decolorizing activity for molasses pigment by *Coriolus versicolor* Ps4a. Agric Biol Chem 1985; 49: 2041-5.

[123] Heinfling A, Ruiz-Dueñas FJ, Martínez MJ, Bergbauer M, Szewzyk U, Martínez AT. A study on reducing substrates of manganese-oxidizing peroxidases from *Pleurotus eryngii* and *Bjerkandera adusta.* FEBS Lett 1998; 428(3): 141-6. [http://dx.doi.org/10.1016/S0014-5793(98)00512-2] [PMID: 9654123]

[124] Linko S. Production of *Phanerochaete chrysosporium* lignin peroxidase. Biotechnol Adv 1992; 10(2): 191-236. [http://dx.doi.org/10.1016/0734-9750(92)90003-R] [PMID: 14544534]

[125] Borel E, Hostettler F, Deuel H. Quantitative zuckerbestimmung mit 3,5-dinitrosalicylsaure und phenol. Helv Chim Acta 1952; 35(1): 115-20. [http://dx.doi.org/10.1002/hlca.19520350116]

[126] Guillen F, Martinez AT, Martinez MJ. Production of hydrogen peroxide by aryl-alcohol oxidase from the ligninolytic fungus *Pleurotus eryngii.* Appl Microbiol Biotechnol 1990; 32: 465-9. [http://dx.doi.org/10.1007/BF00903784]

[127] Guillén F, Martínez AT, Martínez MJ. Substrate specificity and properties of the aryl-alcohol oxidase from the ligninolytic fungus *Pleurotus eryngii.* Eur J Biochem 1992; 209(2): 603-11. [http://dx.doi.org/10.1111/j.1432-1033.1992.tb17326.x] [PMID: 1425667]

[128] Marzullo L, Cannio R, Giardina P, Santini MT, Sannia G. Veratryl alcohol oxidase from *Pleurotus ostreatus* participates in lignin biodegradation and prevents polymerization of laccase-oxidized substrates. J Biol Chem 1995; 270(8): 3823-7. [http://dx.doi.org/10.1074/jbc.270.8.3823] [PMID: 7876125]

[129] Asada Y, Watanabe A, Ohtsu Y, Kuwahara M. Purification and characterization of an arylalcohol oxidase from lignin-degrading basidiomycete *Phanerochaete chrysosporium.* Biosci Biotechnol Biochem 1995; 59(7): 1339-41. [http://dx.doi.org/10.1271/bbb.59.1339]

[130] Sarkar S, Martínez AT, Martínez MJ. Biochemical and molecular characterization of a manganese peroxidase isoenzyme from *Pleurotus ostreatus.* Biochim Biophys Acta Protein Struct Mol Enzymol 1997; 1339(1): 23-30. [http://dx.doi.org/10.1016/S0167-4838(96)00201-4] [PMID: 9165096]

[131] Schneegaß I, Hofrichter M, Scheibner K, Fritsche W. Purification of the main manganese peroxidase isoenzyme MnP2 from the white-rot fungus *Nematoloma frowardii* b19. Appl Microbiol Biotechnol 1997; 48(5): 602-5. [http://dx.doi.org/10.1007/s002530051102]

[132] Moreira PR, Duez C, Dehareng D, *et al.* Molecular characterisation of a versatile peroxidase from a *Bjerkandera* strain. J Biotechnol 2005; 118(4): 339-52. [http://dx.doi.org/10.1016/j.jbiotec.2005.05.014] [PMID: 16026883]

[133] Johansson T, Nyman PO. A cluster of genes encoding major isozymes of lignin peroxidase and manganese peroxidase from the white-rot fungus *Trametes versicolor.* Gene 1996; 170(1): 31-8.

[http://dx.doi.org/10.1016/0378-1119(95)00846-2] [PMID: 8621085]

[134] Lobos S, Larraín J, Salas L, Cullen D, Vicuña R. Isoenzymes of manganese-dependent peroxidase and laccase produced by the lignin-degrading basidiomycete *Ceriporiopsis subvermispora.* Microbiology (Reading) 1994; 140(10): 2691-8.
[http://dx.doi.org/10.1099/00221287-140-10-2691] [PMID: 8000540]

[135] Forrester IT, Grabski AC, Mishra C, *et al.* Characteristics and N-terminal amino acid sequence of a manganese peroxidase purified from *Lentinula edodes* cultures grown on a commercial wood substrate. Appl Microbiol Biotechnol 1990; 33(3): 359-65.
[http://dx.doi.org/10.1007/BF00164536] [PMID: 1366641]

[136] de Boer HA, Zhang YZ, Collins C, Reddy CA. Analysis of nucleotide sequences of two ligninase cDNAs from a white-rot filamentous fungus, *Phanerochaete chrysosporium.* Gene 1987; 60(1): 93-102.
[http://dx.doi.org/10.1016/0378-1119(87)90217-4] [PMID: 3440521]

[137] Claus H, Faber G, König H. Redox-mediated decolorization of synthetic dyes by fungal laccases. Appl Microbiol Biotechnol 2002; 59(6): 672-8.
[http://dx.doi.org/10.1007/s00253-002-1047-z] [PMID: 12226723]

[138] Sugano Y, Matsushima Y, Tsuchiya K, Aoki H, Hirai M, Shoda M. Degradation pathway of an anthraquinone dye catalyzed by a unique peroxidase DyP from *Thanatephorus cucumeris* Dec 1. Biodegradation 2009; 20(3): 433-40.
[http://dx.doi.org/10.1007/s10532-008-9234-y] [PMID: 19009358]

[139] Sugano Y, Sasaki K, Shoda M. cDNA cloning and gene *dyp* from *Geotrichum candidum* Dec 1. J Biosci Bioeng 1999; 87: 411-7.
[http://dx.doi.org/10.1016/S1389-1723(99)80087-5] [PMID: 16232492]

[140] Sambrook J, Fritsch EF, Maniatis T. Molecular cloning: a laboratory manual. 2nd ed., New York: Cold Spring Harbor Laboratory Press 1989.

[141] Poulos TL, Kraut J. The stereochemistry of peroxidase catalysis. J Biol Chem 1980; 255(17): 8199-205.
[http://dx.doi.org/10.1016/S0021-9258(19)70630-9] [PMID: 6251047]

[142] Poulos TL, Edwards SL, Wariishi H, Gold MH. Crystallographic refinement of lignin peroxidase at 2 A. J Biol Chem 1993; 268(6): 4429-40.
[http://dx.doi.org/10.1016/S0021-9258(18)53627-9] [PMID: 8440725]

[143] Sugano Y, Nakano R, Sasaki K, Shoda M. Efficient heterologous expression in *Aspergillus oryzae* of a unique dye-decolorizing peroxidase, DyP, of *Geotrichum candidum* Dec 1. Appl Environ Microbiol 2000; 66(4): 1754-8.
[http://dx.doi.org/10.1128/AEM.66.4.1754-1758.2000] [PMID: 10742277]

[144] Sato T, Hara S, Matsui T, *et al.* A unique dye-decolorizing peroxidase, DyP, from *Thanatephorus cucumeris* Dec 1: heterologous expression, crystallization and preliminary X-ray analysis. Acta Crystallogr D Biol Crystallogr 2004; 60(1): 149-52.
[http://dx.doi.org/10.1107/S0907444903025472] [PMID: 14684913]

[145] Sugano Y. DyP-type peroxidases comprise a novel heme peroxidase family. Cell Mol Life Sci 2009; 66(8): 1387-403.
[http://dx.doi.org/10.1007/s00018-008-8651-8] [PMID: 19099183]

[146] Shakeri M, Sugano Y, Shoda M. Production of dye-decolorizing peroxidase (rDyP) from complex substrates by repeated-batch and fed-batch cultures of recombinant *Aspergillus oryzae.* J Biosci Bioeng 2007; 103(2): 129-34.
[http://dx.doi.org/10.1263/jbb.103.129] [PMID: 17368394]

[147] Ueno S, Miyama M, Ohashi Y, Izumiya M, Kusaka I. Secretory enzyme production and conidiation of *Aspergillus oryzae* in submerged liquid culture. Appl Microbiol Biotechnol 1987; 26(3): 273-6.

[http://dx.doi.org/10.1007/BF00286323]

[148] Bu'Lock JD. Intermediary metabolism and antibiotic synthesis. Adv Appl Microbiol 1961; 3: 293-342. [http://dx.doi.org/10.1016/S0065-2164(08)70514-8] [PMID: 13874183]

[149] Sekiguchi J, Gaucher GM. Conidiogenesis and secondary metabolism in *Penicillium urticae*. Appl Environ Microbiol 1977; 33(1): 147-58. [http://dx.doi.org/10.1128/aem.33.1.147-158.1977] [PMID: 836020]

[150] Biesebeke R, van Biezen N, de Vos WM, van den Hondel CAMJJ, Punt PJ. Different control mechanisms regulate glucoamylase and protease gene transcription in *Aspergillus oryzae* in solid-state and submerged fermentation. Appl Microbiol Biotechnol 2005; 67(1): 75-82. [http://dx.doi.org/10.1007/s00253-004-1807-z] [PMID: 15800731]

[151] Pandy A, Selvakumar P, Soccol RC, Nigma P. Solid state fermentation for the production of industrial enzymes. Curr Sci 1999; 77: 149-61.

[152] Ruckenstein E, Wang XB. Production of lignin peroxidase by *Phanerochaete chrysosporium* immobilized on porous poly(styrene-divinylbenzene) carrier and its application to the degrading of 2-chlorophenol. Biotechnol Bioeng 1994; 44(1): 79-86. [http://dx.doi.org/10.1002/bit.260440112] [PMID: 18618449]

[153] Ziegenhagen D, Hofrichter M. A simple and rapid method to gain high amounts of manganese peroxidase with immobilized mycelium of the agaric white-rot fungus *Clitocybula dusenii*. Appl Microbiol Biotechnol 2000; 53(5): 553-7. [http://dx.doi.org/10.1007/s002530051656] [PMID: 10855715]

[154] Nüske J, Scheibner K, Dornberger U, Ullrich R, Hofrichter M. Large scale production of manganese-peroxidase using agaric white-rot fungi. Enzyme Microb Technol 2002; 30(4): 556-61. [http://dx.doi.org/10.1016/S0141-0229(02)00013-3]

[155] Wang L, Ridgway D, Gu T, Moo-Young M. Bioprocessing strategies to improve heterologous protein production in filamentous fungal fermentations. Biotechnol Adv 2005; 23(2): 115-29. [http://dx.doi.org/10.1016/j.biotechadv.2004.11.001] [PMID: 15694123]

[156] Tung TQ, Miyata N, Iwahori K. Growth of *Aspergillus oryzae* during treatment of cassava starch processing wastewater with high content of suspended solids. J Biosci Bioeng 2004; 97(5): 329-35. [http://dx.doi.org/10.1016/S1389-1723(04)70214-5] [PMID: 16233638]

[157] Hawkes JJ, Limaye MS, Coakley WT. Filtration of bacteria and yeast by ultrasound-enhanced sedimentation. J Appl Microbiol 1997; 82(1): 39-47. [http://dx.doi.org/10.1111/j.1365-2672.1997.tb03295.x] [PMID: 9113876]

[158] Schiraldi C, Marulli F, Di Lernia I, Martino A, De Rosa M. A microfiltration bioreactor to achieve high cell density in *Sulfolobus solfataricus* fermentation. Extremophiles 1999; 3(3): 199-204. [http://dx.doi.org/10.1007/s007920050117] [PMID: 10484176]

[159] Wan HM, Chen CC, Giridhar R, Chang TS, Wu WT. Repeated-batch production of kojic acid in a cell-retention fermenter using *Aspergillus oryzae* M3B9. J Ind Microbiol Biotechnol 2005; 32(6): 227-33. [http://dx.doi.org/10.1007/s10295-005-0230-5] [PMID: 15895266]

[160] Shakeri M, Sugano Y, Shoda M. Stable repeated-batch production of recombinant dye-decolorizing peroxidase (rDyP) from *Aspergillus oryzae*. J Biosci Bioeng 2008; 105(6): 683-6. [http://dx.doi.org/10.1263/jbb.105.683] [PMID: 18640612]

[161] Metz B, Kossen NWF. The growth of molds in the form of pellets–a literature review. Biotechnol Bioeng 1977; 19(6): 781-99. [http://dx.doi.org/10.1002/bit.260190602]

[162] Yesilada O, Cing S, Asma D. Decolourisation of the textile dye Astrazon Red FBL by *Funalia trogii* pellets. Bioresour Technol 2002; 81(2): 155-7. [http://dx.doi.org/10.1016/S0960-8524(01)00117-1] [PMID: 11762908]

[163] Soares GMB, de Amorim MTP, Costa-Ferreira M. Use of laccase together with redox mediators to decolourize Remazol Brilliant Blue R. J Biotechnol 2001; 89(2-3): 123-9. [http://dx.doi.org/10.1016/S0168-1656(01)00302-9] [PMID: 11500205]

[164] Nagai M, Sato T, Watanabe H, Saito K, Kawate M, Enei H. Purification and characterization of an extracellular laccase from the edible mushroomL *Lentinula edodes*, and decolorization of chemically different dyes. Appl Microbiol Biotechnol 2002; 60(3): 327-35. [http://dx.doi.org/10.1007/s00253-002-1109-2] [PMID: 12436315]

[165] Palmieri G, Cennamo G, Sannia G. Remazol Brilliant Blue R decolourisation by the fungus *Pleurotus ostreatus* and its oxidative enzymatic system. Enzyme Microb Technol 2005; 36(1): 17-24. [http://dx.doi.org/10.1016/j.enzmictec.2004.03.026]

[166] Lin YC, Wu JY, Chen TL. Production of *Acinetobacter radioresistens* lipase with repeated batch culture in presence of nonwoven fabric. Biotechnol Bioeng 2001; 76(3): 214-8. [http://dx.doi.org/10.1002/bit.1185] [PMID: 11668456]

[167] Hata Y, Ishida H. Glucoamylase-encoding genes of *Aspergillus oryzae.* Seibutsu-kogaku 2000; 78: 120-7.

[168] Nagel FJJI, Tramper J, Bakker MSN, Rinzema A. Model for on-line moisture-content control during solid-state fermentation. Biotechnol Bioeng 2001; 72(2): 231-43. [http://dx.doi.org/10.1002/1097-0290(20000120)72:2<231::AID-BIT11>3.0.CO;2-S] [PMID: 11114660]

[169] Tengerdy RP. Solid substrate fermentation. Trends Biotechnol 1985; 3(4): 96-9. [http://dx.doi.org/10.1016/0167-7799(85)90092-7]

[170] Hata Y, Ishida H, Kojima Y, *et al.* Comparison of two glucoamylases produced by *Aspergillus oryzae* in solid-state culture (koji) and in submerged culture. J Ferment Bioeng 1997; 84(6): 532-7. [http://dx.doi.org/10.1016/S0922-338X(97)81907-1]

[171] Morita H, Matsunaga M, Mizuno K, Fujio Y. A comparison of raw starch-digesting glucoamylase production in liquid and solid cultures of *Rhizopus strains*. J Gen Appl Microbiol 1998; 44(3): 211-6. [http://dx.doi.org/10.2323/jgam.44.211] [PMID: 12501430]

[172] Hashimoto T, Morishita M, Iwashita K, *et al.* Production and some properties of salt-tolerant β-xylosidases from a Shoyu Koji Mold, *Aspergillus oryzae* in solid and liquid cultures. J Biosci Bioeng 1999; 88(5): 479-83. [http://dx.doi.org/10.1016/S1389-1723(00)87662-8] [PMID: 16232648]

[173] Ishida H, Hata Y, Kawato A, Abe Y, Suginami K, Imayasu S. Identification of functional elements that regulate the glucoamylase-encoding gene (*glaB*) expressed in solid-state culture of *Aspergillus oryzae.* Curr Genet 2000; 37(6): 373-9. [http://dx.doi.org/10.1007/s002940000118] [PMID: 10905427]

[174] Maruyama JI, Ohnuma H, Yoshikawa A, Kadokura H, Nakajima H, Kitamoto K. Production and product quality assessment of human hepatitis B virus pre-S2 antigen in submerged and solid-state cultures of *Aspergillus oryzae.* J Biosci Bioeng 2000; 90(1): 118-20. [http://dx.doi.org/10.1016/S1389-1723(00)80045-6] [PMID: 16232829]

[175] Shimokawa T, Hirai M, Shoda M, Sugano Y. Efficient dye decolorization and production of dye decolorizing enzymes by the basidiomycete *Thanatephorus cucumeris* Dec 1 in a liquid and solid hybrid culture. J Biosci Bioeng 2008; 106(5): 481-7. [http://dx.doi.org/10.1263/jbb.106.481] [PMID: 19111644]

[176] Hocking AD, Pitt JI. Dichloran-glycerol medium for enumeration of xerophilic fungi from low-moisture foods. Appl Environ Microbiol 1980; 39(3): 488-92. [http://dx.doi.org/10.1128/aem.39.3.488-492.1980] [PMID: 7387151]

[177] Zhao J, Koker TH, Janse BJH. Comparative studies of lignin peroxidases and manganese-dependent peroxidases produced by selected white rot fungi in solid media. FEMS Microbiol Lett 1996; 145(3):

393-9.
[http://dx.doi.org/10.1111/j.1574-6968.1996.tb08606.x]

[178] Bellon-Maurel V, Orliac O, Christen P. Sensors and measurements in solid state fermentation: a review. Process Biochem 2003; 38(6): 881-96.
[http://dx.doi.org/10.1016/S0032-9592(02)00093-6]

[179] Gold MH, Alic M. Molecular biology of the lignin-degrading basidiomycete *Phanerochaete chrysosporium.* Microbiol Rev 1993; 57(3): 605-22.
[http://dx.doi.org/10.1128/mr.57.3.605-622.1993] [PMID: 8246842]

[180] Mester T, Field JA. Characterization of a novel manganese peroxidase-lignin peroxidase hybrid isozyme produced by *Bjerkandera* species strain BOS55 in the absence of manganese. J Biol Chem 1998; 273(25): 15412-7.
[http://dx.doi.org/10.1074/jbc.273.25.15412] [PMID: 9624124]

[181] Kamitsuji H, Honda Y, Watanabe T, Kuwahara M. Production and induction of manganese peroxidase isozymes in a white-rot fungus *Pleurotus ostreatus.* Appl Microbiol Biotechnol 2004; 65(3): 287-94.
[http://dx.doi.org/10.1007/s00253-003-1543-9] [PMID: 14767623]

[182] Villena GK, Gutierrez-Correa M. Production of lignocellulolytic enzymes by *Aspergillus niger* biofilms at variable water activities. Electron J Biotechnol 2007; 10: 124-40.
[http://dx.doi.org/10.2225/vol10-issue1-fulltext-2]

[183] Villena GK, Gutiérrez-Correa M. Morphological patterns of *Aspergillus niger* biofilms and pellets related to lignocellulolytic enzyme productivities. Lett Appl Microbiol 2007; 45(3): 231-7.
[http://dx.doi.org/10.1111/j.1472-765X.2007.02183.x] [PMID: 17718832]

[184] Hölker U, Lenz J. Solid-state fermentation — are there any biotechnological advantages? Curr Opin Microbiol 2005; 8(3): 301-6.
[http://dx.doi.org/10.1016/j.mib.2005.04.006] [PMID: 15939353]

[185] Shimokawa T, Shoda M, Sugano Y. Purification and characterization of two DyP isozymes from *Thanatephorus cucumeris* Dec 1 specifically expressed in an air-membrane surface bioreactor. J Biosci Bioeng 2009; 107(2): 113-5.
[http://dx.doi.org/10.1016/j.jbiosc.2008.09.013] [PMID: 19217546]

[186] Shakeri M, Shoda M. Decolorization of an anthraquinone dye by the recombinant dye-decolorizing peroxidase (rDyP) immobilized on mesoporous materials. J Mol Catal, B Enzym 2008; 54(1-2): 42-9.
[http://dx.doi.org/10.1016/j.molcatb.2007.12.009]

[187] Shakeri M, Shoda M. Efficient decolorization of an anthraquinone dye by recombinant dye-decolorizing peroxidase (rDyP) immobilized in silica-based mesocellular foam. J Mol Catal, B Enzym 2010; 62(3-4): 277-81.
[http://dx.doi.org/10.1016/j.molcatb.2009.11.007]

[188] Vinu A, Murugesan V, Hartmann M. Adsorption of lysozyme over mesoporous molecular sieves MCM-41 and SBA-15: Influence of pH and aluminum incorporation. J Phys Chem B 2004; 108(22): 7323-30.
[http://dx.doi.org/10.1021/jp037303a]

[189] Moelans D, Cool P, Baeyens J, Vansant EF. Using mesoporous silica materials to immobilise biocatalysis-enzymes. Catal Commun 2005; 6(4): 307-11.
[http://dx.doi.org/10.1016/j.catcom.2005.02.005]

[190] Díaz JF, Balkus KJ Jr. Enzyme immobilization in MCM-41 molecular sieve. J Mol Catal, B Enzym 1996; 2(2-3): 115-26.
[http://dx.doi.org/10.1016/S1381-1177(96)00017-3]

[191] Han YJ, Watson JT, Stucky GD, Butler A. Catalytic activity of mesoporous silicate-immobilized chloroperoxidase. J Mol Catal, B Enzym 2002; 17(1): 1-8.
[http://dx.doi.org/10.1016/S1381-1177(01)00072-8]

[192] Blanco RM, Terreros P, Fernández-Pérez M, Otero C, Díaz-González G. Functionalization of mesoporous silica for lipase immobilization. J Mol Catal, B Enzym 2004; 30(2): 83-93. [http://dx.doi.org/10.1016/j.molcatb.2004.03.012]

[193] Hernández-Ruiz J, Arnao MB, Hiner ANP, García-Cánovas F, Acosta M. Catalase-like activity of horseradish peroxidase: relationship to enzyme inactivation by H_2O_2. Biochem J 2001; 354(1): 107-14. [http://dx.doi.org/10.1042/bj3540107] [PMID: 11171085]

[194] González-Sáiz JM, Pizarro C. Polyacrylamide gels as support for enzyme immobilization by entrapment. Effect of polyelectrolyte carrier, pH and temperature on enzyme action and kinetics parameters. Eur Polym J 2001; 37(3): 435-44. [http://dx.doi.org/10.1016/S0014-3057(00)00151-8]

[195] Di Martino S, El-Sheriff H, Diano N, *et al.* Urea removal from agricultural waste waters by means of urease immobilized on nylon membranes grafted with cyclohexyl-methacrylate. Appl Catal B 2003; 46(3): 613-29. [http://dx.doi.org/10.1016/S0926-3373(03)00323-0]

[196] Palma C, Moreira MT, Feijoo G, Lema JM. Enhanced catalytic properties of MnP by exogenous addition of manganese and hydrogen peroxide. Biotechnol Lett 1997; 19(3): 263-8. [http://dx.doi.org/10.1023/A:1018313825723]

[197] Hiner ANP, Rodríguez-López JN, Arnao MB, Raven ELLOYD, García-Cánovas F, Acosta M. Kinetic study of the inactivation of ascorbate peroxidase by hydrogen peroxide. Biochem J 2000; 348(2): 321-8. [http://dx.doi.org/10.1042/bj3480321] [PMID: 10816425]

[198] Arnao MB, Acosta M, del Río JA, García-Cánovas F. Inactivation of peroxidase by hydrogen peroxide and its protection by a reductant agent. Biochim Biophys Acta Protein Struct Mol Enzymol 1990; 1038(1): 85-9. [http://dx.doi.org/10.1016/0167-4838(90)90014-7] [PMID: 2317519]

[199] Mielgo I, López C, Moreira MT, Feijoo G, Lema JM. Oxidative degradation of azo dyes by manganese peroxidase under optimized conditions. Biotechnol Prog 2003; 19(2): 325-31. [http://dx.doi.org/10.1021/bp020136w] [PMID: 12675567]

[200] Zhang G, Nicell JA. Treatment of aqueous pentachlorophenol by 2 horseradish peroxidase and hydrogen peroxide. Water Res 2000; 34: 1629-1637. [http://dx.doi.org/10.1016/S0043-1354(99)00326-7]

[201] Lopez C, Moreira M, Feijoo G, Lema JM. Dye decolorization by manganese peroxidase in an enzymatic membrane bioreactor. Biotechnol Prog 2004; 20:74-81. [http://dx.doi.org/10.1021/bp030025c] [PMID: 14763826]

[202] Kasinath A, Novotny C, Svobodova K, Patel KC, Sasek V. Decolorization of synthetic dyes by *Lrpex lacteus* in liquid culture and packed-bed bioreactor. Enzyme Microb Technol 2003; 32:167-173. [http://dx.doi.org/10.1016/S0141-0229(02)00279-X]

[203] Trupkin S, Levin L, Forchiassin F, Viale A. Optimization of a culture medium for ligninolytic enzyme production and synthetic dye decolorization using response surface methodology. J Ind Microbiol Technol 2003; 30: 682-690. [http://dx.doi.org/10.1007/s10295-003-0099-0] [PMID: 14648345]

[204] Champagne PP, Ramsay JA. Contribution of manganese peroxidase and laccase to dye decolorization by *Trametes versicolor*. Appl Microbiol Biotechnol 2005; 69: 276-285. [http://dx.doi.org/10.1007/s00253-005-1964-8] [PMID: 15834615]

[205] Zille A, Tzanov T, Gubitz GM, Cavaco-Paulo A. Immobilized laccase for decolorization of reactive black 5 dyeing effluents. Biotech Lett 2003; 25: 1473-1477. [http://dx.doi.org/10.1023/A:1025032323517] [PMID: 14514053]

[206] Torres E, Bustos-Jaimes I, Le Borgne S. Potential use of oxidative enzymes for the detoxification of

organic pollutants. Appl Catal B 2003; 46(1): 1-15.
[http://dx.doi.org/10.1016/S0926-3373(03)00228-5]

[207] Mechichi T, Mhiri N, Sayadi S. Remazol Brilliant Blue R decolourization by the laccase from *Trametes trogii.* Chemosphere 2006; 64(6): 998-1005.
[http://dx.doi.org/10.1016/j.chemosphere.2005.12.061] [PMID: 16473395]

[208] Ohmomo S, Aoshima I, Tozawa Y, Sakurada N, Ueda K. Purification and some properties of melanoidin decolorizing enzymes, P-111 and P-IV, from mycelia of *Coriolus versicolor* Ps4a. Agric Biol Chem 1985; 49: 2047-53.

[209] Dehorter B, Blondeau R. Isolation of an extracellular Mn-dependent enzyme mineralizing melanoidins from the white rot fungus *Trametes versicolor.* FEMS Microbiol Lett 1993; 109(1): 117-22.
[http://dx.doi.org/10.1111/j.1574-6968.1993.tb06153.x]

[210] Paszczynski A, Pasti MB, Goszczynski S, Crawford DL, Crawford RL. New approach to improve degradation of recalcitrant azo dyes by *Streptomyces* spp. and *Phanerochaete chrysosporium.* Enzyme Microb Technol 1991; 13(5): 378-84.
[http://dx.doi.org/10.1016/0141-0229(91)90198-J]

[211] Kirkpatrick N, Palmer JM. Semi-continuous ligninase production using foam-immobilised *Phanerochaete chrysosporium.* Appl Microbiol Biotechnol 1987; 27(2): 129-33.
[http://dx.doi.org/10.1007/BF00251934]

[212] Capdevila C, Corrieu G, Asther M. A feed-harvest culturing method to improve lignin peroxidase production by *Phanerochaete chrysosporium* INA-12 immobilized on polyurethane foam. J Ferment Bioeng 1989; 68(1): 60-3.
[http://dx.doi.org/10.1016/0922-338X(89)90217-1]

[213] Lee TH, Chun GT, Chang YK. Development of sporulation/immobilization method and its application for the continuous production of cyclosporin A by *Tolypocladium inflatum.* Biotechnol Prog 1997; 13 (5): 546-50.
[http://dx.doi.org/10.1021/bp970069j] [PMID: 9376111]

[214] Yang FC, Yu JT. Development of a bioreactor system using an immobilized white rot fungus for decolorization. Bioprocess Eng 1996; 15(6): 307-10.
[http://dx.doi.org/10.1007/BF02426439]

[215] Yang FC, Yu JT. Development of a bioreactor system using an immobilized white rot fungus for decolorization. Bioprocess Eng 1996; 16(1): 9-11.
[http://dx.doi.org/10.1007/s004490050280]

[216] Shintani N, Sugano Y, Shoda M. Decolorization of kraft pulp bleaching effluent by a newly isolated fungus, *Geotrichum candidum* Dec 1. J Wood Sci 2002; 48(5): 402-8.
[http://dx.doi.org/10.1007/BF00770700]

[217] Jaspers CJ, Penninckx MJ. Adsorption effects in the decolorization of a Kraft bleach plant effluent by *Phanerochaete chrysosporium.* Biotechnol Lett 1996; 18(11): 1257-60.
[http://dx.doi.org/10.1007/BF00129950]

[218] Bourbonnais R, Paice MG. Oxidation of non-phenolic substrates. FEBS Lett 1990; 267(1): 99-102.
[http://dx.doi.org/10.1016/0014-5793(90)80298-W] [PMID: 2365094]

[219] Gerkov M, Priha M, Talka E, Valttila O. Chlorinated organic compounds in effluents treatment at kraft mills. Tappi J 1988; 71: 175-84.

[220] Lee SH, Kondo R, Sakai K. Treatment of kraft bleaching effluents by lignin degrading fungi. III. Treatment by newly found fungus KS-62 without additional nutrients. Mokuzai Gakkaishi 1994; 40: 612-9.

[221] Momohara I, Matsumoto Y, Ishizu Y, Chang HM. Decolorization mechanism of spent bleaching liquor by *Phanerochaete chrysosporium.* Proceedings of the 5th ISPWC. 373-6.

[222] Michel FC Jr, Dass SB, Grulke EA, Reddy CA. Role of manganese peroxidases and lignin peroxidases of *Phanerochaete chrysosporium* in the decolorization of kraft bleach plant effluent. Appl Environ Microbiol 1991; 57(8): 2368-75.
[http://dx.doi.org/10.1128/aem.57.8.2368-2375.1991] [PMID: 1768105]

[223] Gold MH, Wriishi H, Valli K. Extracellular peroxidases involved in lignin degradation by the white rot basidiomycete *Phanerochaete chrysosporium*. In: Whitaker JR, PE Sonnet, Eds. Biocatalysis in agricultural biotechnology. Washington. DC: America Chemical Society 1989; pp. 127-36.

[224] Lackner R, Srebotnik E, Messner K. Oxidative degradation of high molecular weight chlorolignin by manganese peroxidase of *Phanerochaete chrysosporium*. Biochem Biophys Res Commun 1991; 178(3): 1092-8.
[http://dx.doi.org/10.1016/0006-291X(91)91004-V] [PMID: 1872832]

[225] Shintani N, Shoda M. Decolorization of oxygen-delignified bleaching effluent and biobleaching of oxygen-delignified kraft pulp by non-white-rot fungus *Geotrichum candidum* Dec 1. J Environ Sci 2013; 25 (Suppl. 1): S164-8.
[http://dx.doi.org/10.1016/S1001-0742(14)60649-5] [PMID: 25078823]

[226] Katagiri N, Tsutsumi Y, Nishida T. Correlation of brightening with cumulative enzyme activity related to lignin biodegradation during biobleaching of kraft pulp by white rot fungi in the solid-state fermentation system. Appl Environ Microbiol 1995; 61(2): 617-22.
[http://dx.doi.org/10.1128/aem.61.2.617-622.1995] [PMID: 7574600]

SUBJECT INDEX

Makoto Shoda
All rights reserved-© 2024 Bentham Science Publishers

C

D

I

L

M

N

O

V

W

X

Made in United States
Orlando, FL
04 November 2026

90206099R00100